Studies in Computational Intelligence

Volume 736

Series editor

Janusz Kacprzyk, Polish Academy of Sciences, Warsaw, Poland
e-mail: kacprzyk@ibspan.waw.pl

About this Series

The series "Studies in Computational Intelligence" (SCI) publishes new developments and advances in the various areas of computational intelligence—quickly and with a high quality. The intent is to cover the theory, applications, and design methods of computational intelligence, as embedded in the fields of engineering, computer science, physics and life sciences, as well as the methodologies behind them. The series contains monographs, lecture notes and edited volumes in computational intelligence spanning the areas of neural networks, connectionist systems, genetic algorithms, evolutionary computation, artificial intelligence, cellular automata, self-organizing systems, soft computing, fuzzy systems, and hybrid intelligent systems. Of particular value to both the contributors and the readership are the short publication timeframe and the worldwide distribution, which enable both wide and rapid dissemination of research output.

More information about this series at http://www.springer.com/series/7092

Boris Kryzhanovsky · Witali Dunin-Barkowski
Vladimir Redko
Editors

Advances in Neural Computation, Machine Learning, and Cognitive Research

Selected Papers from the XIX
International Conference on
Neuroinformatics, October 2–6, 2017,
Moscow, Russia

 Springer

Editors
Boris Kryzhanovsky
Scientific Research Institute for System
  Analysis
Russian Academy of Sciences
Moscow
Russia

Vladimir Redko
Scientific Research Institute for System
  Analysis
Russian Academy of Sciences
Moscow
Russia

Witali Dunin-Barkowski
Scientific Research Institute for System
  Analysis
Russian Academy of Sciences
Moscow
Russia

ISSN 1860-949X   ISSN 1860-9503   (electronic)
Studies in Computational Intelligence
ISBN 978-3-319-88283-3   ISBN 978-3-319-66604-4   (eBook)
DOI 10.1007/978-3-319-66604-4

Printed on acid-free paper

This Springer imprint is published by Springer Nature
The registered company is Springer International Publishing AG
The registered company address is: Gewerbestrasse 11, 6330 Cham, Switzerland

Preface

The international conference "Neuroinformatics" is the annual multidisciplinary scientific forum dedicated to the theory and applications of artificial neural networks, the problems of neuroscience and biophysics systems, adaptive behavior, and cognitive studies.

The scope of the conference is wide, ranging from theory of artificial neural networks, machine learning algorithms, and evolutionary programming to neuroimaging and neurobiology.

Main topics of the conference cover theoretical and applied research from the following fields:

neurobiology and neurobionics: cognitive studies, neural excitability, cellular mechanisms, cognition and behavior, learning and memory, motivation and emotion, bioinformatics, computation, modeling, and simulation;

neural networks: neurocomputing and learning, architectures, biological foundations, computational neuroscience, neurodynamics, neuroinformatics, deep learning networks;

machine learning: pattern recognition, Bayesian networks, kernel methods, generative models, information theoretic learning, reinforcement learning, relational learning, dynamical models, classification and clustering algorithms, self-organizing systems;

applications: medicine, signal processing, control, simulation, robotics, hardware implementations, security, finance and business, data mining, natural language processing, image processing, and computer vision.

A total of about 90 reports were presented at the Neuroinformatics-2017 Conference. Of these, 28 papers were selected for which articles were prepared and published in this volume.

Boris Kryzhanovsky
Witali Dunin-Barkowski
Vladimir Redko

Organization

Editorial Board

Boris Kryzhanovsky	Scientific Research Institute for System Analysis of Russian Academy of Sciences
Witali Dunin-Barkowsky	Scientific Research Institute for System Analysis of Russian Academy of Sciences
Vladimir Redko	Scientific Research Institute for System Analysis of Russian Academy of Sciences

Advisory Board

Prof. Alexander N. Gorban (Chair of the International Advisory Board),
Department of Mathematics
University of Leicester
Leicester LE1 7RH
UK
Tel. +44 116 223 14 33
E-mail: ag153@le.ac.uk
Homepage: http://www.math.le.ac.uk/people/ag153/homepage/
Google scholar profile:
http://scholar.google.co.uk/citations?user=D8XkcCIAAAAJ&hl=en

Prof. Nicola Kasabov
Professor of Computer Science and Director KEDRI
Phone: +64 9 921 9506
Email: nkasabov@aut.ac.nz
http://www.kedri.info

Physical Address:
KEDRI
Auckland University of Technology
AUT Tower, Level 7
Corner Rutland and Wakefield Street
Auckland
New Zealand

Postal Address:
KEDRI
Auckland University of Technology
Private Bag 92006
Auckland 1142
New Zealand

Prof. Jun Wang, PhD, FIEEE, FIAPR
Chair Professor of Computational Intelligence
Department of Computer Science
City University of Hong Kong
Kowloon Tong, Kowloon, Hong Kong
Tel. +852 34429701
Fax: +852-34420503
Email: jwang.cs@cityu.edu.hk

Program Committee of the XIX International Conference "Neuroinformatics-2017"

General Chair

Kryzhanovskiy Boris   Scientific Research Institute for System Analysis, Moscow

Co-chairs

Dunin-Barkowski Witali   Scientific Research Institute for System Analysis, Moscow

Gorban Alexander   University of Leicester, Great Britain

Redko Vladimir   Scientific Research Institute for System Analysis, Moscow

Program Committee

Abraham Ajith   Machine Intelligence Research Labs (MIR Labs), Scientific Network for Innovation and Research Excellence, Washington, USA

Anokhin Konstantin   National Research Centre "Kurchatov Institute," Moscow

Baidyk Tatiana   The National Autonomous University of Mexico, Mexico

Balaban Pavel   Institute of Higher Nervous Activity and Neurophysiology of RAS, Moscow

Borisyuk Roman   Plymouth University, UK

Burtsev Mikhail   National Research Centre "Kurchatov Institute," Moscow

Contents

Neural Network Theory

The Analysis of Regularization in Deep Neural Networks Using Metagraph Approach

Yuriy S. Fedorenko, Yuriy E. Gapanyuk$^{(\boxtimes)}$,
and Svetlana V. Minakova

Bauman Moscow State Technical University,
Baumanskaya 2-ya, 5, 105005 Moscow, Russia
{fedyural1235, morgana_93}@mail.ru, gapyu@bmstu.ru

Abstract. The article deals with the overfitting problem in deep neural networks. Finding the model with proper number of parameters matching the simulated process can be a difficult task. There are a range of recommendations how to chose the number of neurons in hidden layers, but most of them don't always work well in practice. As a result, neural networks work in underfitting or overfitting regime. Therefore in practice complex model is usually chosen and regularization strategies are applied. In this paper, the main regularization techniques for multilayer perceptrons including early stopping and dropout are discussed. Regularization representation using metagraph approach is described. In the creation mode, the metagraph representation of the neural network is created using metagraph agents. In the training mode, the training metagraph is created. Thus, different regularization strategies may be embedded into the training algorithm. The special metagraph agent for dropout strategy is developed. Comparison of different regularization techniques is conducted on CoverType dataset. Results of experiments are analyzed. Advantages of early stopping and dropout regularization strategies are discussed.

Keywords: Deep neural network · Regularization · Overfitting · Early stopping · Dropout · Metagraph · Metagraph agent

1 Introduction

The choice of deep models has an essential advantage: it often means a priori proposition that the function, obtained after the models training, can be presented as a composition of simple functions [1]. But at the same time it's important that the number of trained parameters (model complexity) correspond to the solving task. If the model complexity is too low, neural network can't be trained to necessary function (for example, a single layer perceptron can't solve the XOR problem). If the model complexity is too high, the model will be trained not only to necessary dependencies but also to noise. It's clear that in practice it is quite difficult to choose a proper size of model. Therefore, the initial model complexity is usually rather high and you need to use regularization techniques to prevent overfitting.

© Springer International Publishing AG 2018
B. Kryzhanovsky et al. (eds.), *Advances in Neural Computation, Machine Learning, and Cognitive Research*, Studies in Computational Intelligence 736,
DOI 10.1007/978-3-319-66604-4_1

2 Description of the Used Neural Network

For researching regularization strategies we apply multilayer perceptron with three hidden layers. The output and hidden layers of neural network have softmax and relu activation functions accordingly. Traditionally the optimal parameters of model are searched by minimization of negative log-likelihood loss function [2]. Mini-batch SGD with gradient calculation by backpropagation [3] was used for model training. The initial number of neurons in each hidden layer is selected in accordance with the requirement of maintaining balance of the networks width and depth. This allows neural network to reach optimal performance [4]. A smooth reducing of data dimensionality from n inputs to m outputs could be provided using the following expression:

$$\frac{m}{h_k} \leq \frac{h_k}{h_{k-1}} \leq \frac{h_{k-1}}{h_{k-2}} \leq \ldots \leq \frac{h_2}{h_1} \leq \frac{h_1}{n} \, ,$$

Where h_k – is a number of neurons in k-th hidden layer of network. The minimum initial number of neurons of the last hidden layer of the network is $h_{k\,min} = m$.

The initial values of weights are of particular importance, because they influence not only on algorithm convergence but also on it generalization ability [5]. In this research the initial weights are sampled from the uniform distribution with limits:

$$W_{i,j} \sim U\left(-\sqrt{\frac{12}{h_i + h_j}}, \sqrt{\frac{12}{h_i + h_j}}\right),$$

where h_i, h_j is a number of units in (i-1) and i-th layer accordingly.

3 Regularization of Deep Neural Networks

Usually training of the deep model is in three main scenarios:

- 1. The model cannot handle real data, which corresponds to underfitting;
- 2. Model handles real data correctly;
- 3. Model handles not only relevant data, but also many other processes – noises. It's the overfitting regime.

In practice, it is often difficult to find model with a number of parameters, providing the second scenario. Therefore, to obtain the necessary mode, you need to train a complex model and apply regularization to it.

One of the most well-known regularization strategies is L^2 regularization in which the sum of squares of parameters values is added to the loss function. When using L^2 regularization it is observed the greatest reducing of the components from which value of objective function depends less significantly [1]. Also, there is a L^1 regularization in which the sum of modules of parameters values is added to the loss function. It leads to the sparsity of the coefficients in the model. Therefore, this strategy is used by some feature selection algorithms [6].

Recently the early stopping regularization method has become popular. Essentially, when using this method, the model training is stopped when validation set error begins to increase. It may be shown [7] that for many loss functions with simple gradient descent optimization early stopping is equivalent to L^2 regularization, but it automatically determines the correct amount of regularization while weight decay requires many training experiments to select necessary coefficient. So, the early stopping reduces computational cost of training procedure.

Dropout is another one regularization strategy. The using of ensemble of models often leads to improvement of quality of machine learning algorithms. But at the same time the simultaneous training of several models may be computationally intensive. Dropout [8] trains the ensemble of neural networks in a computationally inexpensive way. When using bagging k different models are trained on k different datasets constructed from the original training set. When using dropout, the randomly sampled binary mask that switches off part of the neurons is applied for each input example (mask is formed independently for each example). The probability of including neuron in this mask is a model hyperparameter.

4 Regularization Representation Using Metagraph Approach

In our paper [9] the metagraph representation of perceptron neural network was proposed. The distinguishing feature of this approach is the ability to change the structure of neural network during creation or training using metagraph agents. Metagraph is a kind of complex network model: $MG = \langle V, MV, E \rangle$, where MG – metagraph; V – set of metagraph vertices; MV – set of metagraph metavertices; E – set of metagraph edges. From the general system theory point of view metavertex is a special case of manifestation of emergence principle which means that metavertex with its private attributes and connections became whole that cannot be separated into its component parts.

For metagraph transformation the metagraph agents are used. There are two kinds of metagraph agents: the metagraph function agent ag^F and the metagraph rule agent ag^R. The metagraph function agent serves as function with input and output parameter in form of metagraph: $ag^F = \langle MG_{IN}, MG_{OUT}, AST \rangle$, where ag^F – metagraph function agent; MG_{IN} – input parameter metagraph; MG_{OUT} – output parameter metagraph; AST – abstract syntax tree of metagraph function agent in form of metagraph.

The metagraph rule agent uses rule-based approach: $ag^R = \langle MG, R, AG^{ST} \rangle, R = \{r_i\}, r_i : MG_j \rightarrow OP^{MG}$, where ag^R – metagraph rule agent; MG – working metagraph, a metagraph on the basis of which the rules of agent are performed; R – set of rules r_i; AG^{ST} – start condition (metagraph fragment for start rule check or start rule); MG_j – a metagraph fragment on the basis of which the rule is performed; OP^{MG} – set of actions performed on metagraph.

In this paper, we also use the "active metagraph" concept, which means combination of data metagraph with attached metagraph agent.

The structure of metagraph transformations is presented at Fig. 1.

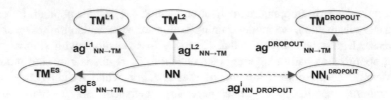

Fig. 1. The structure of metagraph transformations

In the creation mode, the metagraph representation of neural network is created using metagraph agents. According to modelling tasks the complexity of created neuron structure can be various. In the simplest case, the neuron may be considered as a node with activation function. In more complex cases the neuron may be represented as a nested metagraph, which contains metavertices with complex activation function addressing neuron structure. Thus, in the end of creation mode the "Neural Network" (NN) structure is created. In case of deep network this is a flat graph of nodes (neurons) connected with edges. But node may be represented as complex metavertex and neurons of each layer of network may also be combined into metavertex.

In the training mode, the "Training Metagraph" (TM) is created. TM structure is isomorphic to the NN structure. For each node NN_i^n in NN the corresponding metavertex TM_i^n in TM is created. And for each edge NN_i^e in NN the corresponding edge TM_i^e in TM is created. For the TM creation, the agent $ag_{NN \to TM}$ is used. This agent is kind of function agent.

TM may be considered as an active metagraph with metagraph agent ag_{TM} bound to TM graph structure. Agent ag_{TM} implements a particular training algorithm. As a result of training the changed weights are written to the TM_i^n.

Agent ag_{TM} is also created with the $ag_{NN \to TM}$ agent. Different regularization strategies could be embedded into the ag_{TM} training algorithm.

For the single NN we can create several TM with different regularization strategies. For example, TM^{L1} means that $ag_{NN \to TM}$ agent creates TM graph structure with ag_{TM} agent that implements training algorithm with L^1 regularization. Similarly, TM^{L2} stands for L^2 regularization and TM^{ES} stands for early stopping.

It should be noted that neither of these regularization strategies require changing the network structure during the training process. But in case of dropout we have to change the network structure. In this case, the agent $ag_{NN\_DROPOUT}$ is used. This agent is kind of metagraph rule agent.

Using $ag_{NN\_DROPOUT}$ metagraph agent we can implement different dropout strategies. Applying different $ag_{NN\_DROPOUT}^i$ agents to the original NN structure we resulting the set of modified $NN_i^{DROPOUT}$ network structures. For each $NN_i^{DROPOUT}$ the corresponding $TM^{DROPOUT}$ is created for network training.

At Fig. 1 the transformation with structure changing is shown with dashed arrow, the transformation without structure changing is shown with solid arrow.

Thus, the metagraph approach allows representing neural network training with different regularization strategies either with or without network structure transformation.

5 Experiments

The experimental analysis was conducted on CoverType dataset [10]. It contains 15,120 examples in training set (11,340 examples for training set and 3,780 for validation set) and 565,892 examples in test set. Series of experiments with L^1, L^2 regularization, early stopping and dropout was conducted. The results of experiments are presented at Fig. 2 and Table 1. On each graph the loss function value on training and test set is presented. The dashed lines correspond to experiments with early stopping. The results of experiments show that L^1 regularization works slightly worse than L^2 regularization. It's not surprising because L^1 regularization leads to sparse representation sacrificed prediction quality. L^2 regularization aims to search better solution, but it does not aim to provide sparse representation or model simplicity. Using early stopping without regularization allows obtaining reasonably good result without computationally expensive selection of parameters. Dropout shows the best results with less training epoch than L^2 regularization. So with dropout network is trained better and faster than when using other methods.

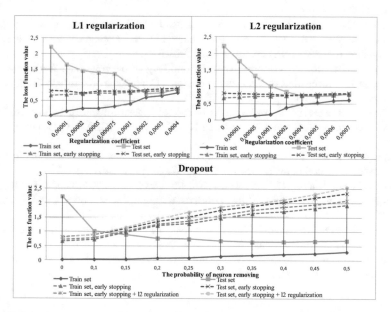

Fig. 2. Training perceptron with three hidden layers on CoverType dataset with different regularization strategies

Table 1. The best error rate of classification for different regularization techniques

Dataset/Regularization type	Without regularization	L^1	L^2	Dropout
CoverType	24.84%	24.59%	24.58%	22.39%

6 Conclusion

In real tasks the models with relatively large number of parameters are usually chosen and regularization techniques are used to reduce overfitting in these models. Along with L^1 and L^2 regularizations there are more effective techniques such as early stopping and dropout. Early stopping automatically determines the correct value of regularization coefficient, so it less computationally expensive than L^2 regularization. Experimental results show that the most effective regularization strategy is dropout. It reduces overall training time of the model as well as error rate.

References

1. Goodfellow, I., Bengio, Y., Courvile, A.: Deep Learning. MIT Press, Cambridge (2016). 787 p
2. Bishop, C.: Pattern Recognition and Machine Learning. Springer, New York (2006). 758 p
3. Haykin, S.: Neural Networks: A Comprehensive Foundation, 2nd edn. Prentice Hall, New Jersey (1999). 1056 p
4. Szegedy, C., Vanhoucke, V., Ioffe, S., Shlens, J., Wojna, Z.: Rethinking the inception architecture for computer vision. In: Conference on Computer Vision and Pattern Recognition (CVPR), USA, pp. 2818–2826 (2016)
5. Glorot, X., Bengio, Y.: Understanding the difficulty of training deep feedforward neural networks. In: Proceedings of the 13th International Conference on Artificial Intelligence and Statistics, Sardinia, Italy, pp. 249–256 (2010)
6. Bishop, C.: Regularization and complexity control in feed-forward networks. In: Proceedings International Conference on Artificial Neural Networks ICANN 1995, vol. 1, pp. 141–148
7. Sjoberg, J., Ljung, L.: Overtraining, regularization and searching for a minimum, with application to neural networks. Int. J. Control **62**(6), 1391–1407 (1995)
8. Srivastava, N., Hinton, G., Krizhevsky, A., Sutskever, I., Salakhutdinov, R.: Dropout: a simple way to prevent neural networks from overfitting. J. Mach. Learn. Res. **15**, 1929–1958 (2014)
9. Fedorenko, YuS, Gapanyuk, YuE: Multilevel neural net adaptive models using the metagraph approach. Opt. Mem. Neural Netw. **25**(4), 228–235 (2016)
10. Blackard, J., Dean, J., Anderson, W.: Forest CoverType Dataset. https://archive.ics.uci.edu/ml/datasets/Covertype. Accessed 14 Jun 2017

Adding Noise During Training as a Method to Increase Resilience of Neural Network Solution of Inverse Problems: Test on the Data of Magnetotelluric Sounding Problem

Igor Isaev and Sergey Dolenko[✉]

D.V. Skobeltsyn Institute of Nuclear Physics, M.V. Lomonosov Moscow State University, Moscow, Russia
dolenko@srd.sinp.msu.ru

Abstract. In their previous studies, the authors proposed to use the approach associated with adding noise to the training set when training multilayer perceptron type neural networks to solve inverse problems. For a model inverse problem it was shown that this allows increasing the resilience of neural network solution to noise in the input data with different distributions and various intensity of noise. In the present study, the observed effect was confirmed on the data of the problem of magnetotelluric sounding. Also, maximum noise resilience (maximum quality of the solution) is generally achieved when the level of the noise in the training data set coincides with the level of noise during network application (in the test dataset). Thus, increasing noise resilience of a network when noise is added during its training is associated with the fundamental properties of multilayer perceptron neural networks and not with the properties of the data. So this method can be used solving other multi-parameter inverse problems.

Keywords: Neural networks · Inverse problems · Noise resilience · Training with noise · Regularization

1 Introduction

The inverse problems (IPs) represent a very important class of problems. Almost any task of indirect measurements can be attributed to them. Inverse problems include multiple problems from the areas of geophysics [1], spectroscopy [2], various types of tomography [3], and many others.

Among them is also the IP of magnetotelluric sounding (MTS), whose purpose is the restoration of the distribution of electrical conductivity in the thick of the earth by the values of the components of electromagnetic fields induced by natural sources,

This study has been performed at the expense of the grant of Russian Science Foundation (project no. 14-11-00579).

© Springer International Publishing AG 2018
B. Kryzhanovsky et al. (eds.), *Advances in Neural Computation, Machine Learning, and Cognitive Research*, Studies in Computational Intelligence 736,
DOI 10.1007/978-3-319-66604-4_2

measured on the surface [4]. Due to the shielding of the underlying layers by the over-lying ones, the contribution of the more deep-lying layers to the changes in the parameters of the measured field is smaller.

In the general case, such problems do have no analytical solutions. So usually they are solved by optimization methods based on the repeated solution of the direct problem with the minimization of the residual in the space of observable quantities [1], or by matrix methods using regularization by Tikhonov [5].

Optimization methods have some disadvantages such as high computational cost, the need for good first approximation and, most importantly, the need for a correct model for solving the direct problem. For methods based on regularization, the main problem is the necessity of choosing the regularization parameter. In this paper we consider artificial neural networks (ANN) as an alternative method of solution of various IPs [6–8] free from these drawbacks.

Another feature of IPs is the possible instability of the solution. Therefore, if the input data contains noise, it reduces the effectiveness of the IP solution methods, both traditional and neural network ones. Practical IPs almost always contain noise due to measurement uncertainty. Despite the fact that the neural network itself is resistant to noise, this is often insufficient when solving inverse problems, because the incorrectness of the problem turns out to be more significant than the ability of the network to overcome it. Development of approaches to increase the resilience of IP solution methods to noise is an urgent task.

The method of adding noise to the input data of a neural network when training has been known for a long time. In [9, 10] it was shown to increase the generalization ability of the network. Most often it is considered as one of the ways to avoid neural network overtraining [11–13]. In [14] it is shown that using this method is equivalent to regularization by Tikhonov. In [15] it was demonstrated that using this method also reduced the network training time. Currently, this method is also used when training deep neural networks [16].

One of the authors has previously used this method with neural network solution of IPs of laser spectroscopy with a small number of determined parameters (up to 3) [17–19]. It has been shown that adding noise when training neural networks allowed one to improve the resilience of the IP solution to the noise in the input data; however, the effectiveness of this technique heavily depends on the specifics of the particular task.

The present study is a direct continuation of the study [20], which dealt with the application of the method of adding noise to the training of a perceptron to enhance the resilience of neural network solutions to noise in the input data for a strongly non-linear multi-parameter model IP, where the dependence of the "observable data" from the "parameters" was explicitly set in the form of a polynomial dependence. This IP by its external manifestations is a simplified model analogue of the IP of MTS [4, 21, 22]. In [20], we confirmed the findings of the previous studies on the acceleration of ANN learning and on improving their resilience to noise, and we determined the optimal parameters of this approach.

The aim of the present study was to check the listed effects on the data of the IP of MTS.

2   The Initial Data and the Statement of the IP of MTS

The study was carried out on simulated data obtained through numerical solution of the direct two-dimensional problem of MTS [21]. The calculated "observable values" are Ex, Ey, Hx, Hy components of the induced electromagnetic fields at various frequencies and in various locations (pickets) on the surface of the earth.

The calculation largely depends on the scheme of the parameterization, i.e. on the way of describing the distribution of electrical conductivity (EC) of the underground area. In the present study we use the data obtained for the most general parameterization scheme G_0, which describes the distribution of EC by the values in the corners of rectangular blocks arranged in layers (nodes of a grid), with subsequent interpolation between nodes. The dataset consisted of 12,000 patterns, obtained for random combinations of EC in different nodes in the range from 10^{-4} to 1 Sm/m.

The described data array was used to solve the IP of determining the values of EC in the nodes (output dimensionality of the problem $N_O = 336$ parameters) by the field values (input dimensionality of the problem $N_I = 4$ field components $\times$ 13 frequencies $\times$ 126 pickets = 6552 input features). The solution of this IP is connected with considerable difficulties associated with the large input and output dimensionality of the data, which involves the use of special approaches to its reduction.

The study was conducted for nodes lying vertically one above the other.

3   Description of the Noise

Two types of noise were considered: additive and multiplicative, and two kinds of statistics: uniform noise (uniform distribution) and Gaussian noise (normal distribution). The value of each input feature x_i was transformed in the following way:

$$x_i^{agn} = x_i + norm\_inv(random, \mu = 0, \sigma = noise\_level) \cdot \max(x_i)$$
$$x_i^{aun} = x_i \cdot (1 - 2 \cdot random) \cdot noise\_level \cdot \max(x_i)$$
$$x_i^{mgn} = x_i \cdot (1 + norm\_inv(random, \mu = 0, \sigma = noise\_level))$$
$$x_i^{mun} = x_i \cdot (1 + (1 - 2 \cdot random) \cdot noise\_level)$$

for additive Gaussian (agn), additive uniform (aun), multiplicative Gaussian (mgn) and multiplicative uniform (mun) noise, respectively.

Here, *random* is a random value in the range from 0 to 1, *norm_inv* function returns the inverse normal distribution, *max(x_i)* is the maximum value of a given feature over all patterns, *noise_level* is the level of noise (considered values: 1%, 3%, 5%, 10%, 20%).

To generate noisy data sets (with different types and various levels of noise), based on each pattern from the source sets, 10 samples with different realizations of noise were created. The original datasets (without noise) contained 3000 patterns in the training set, 6000 patterns in the validation set, and 3000 patterns in the test dataset. Thus, each of the noisy training sets and each of the noisy test sets contained 30,000 patterns. No noise was introduced into the validation set (see below).

4   The Use of Artificial Neural Networks

ANNs were used in the following way. Training was performed on the training set of data. To prevent overtraining, the validation set was used (the training was stopped after 500 epochs with no average error reduction on the validation set). An independent evaluation of the results was carried out using test sets.

In our previous study [20] it was shown that the best training option, which demonstrated greater accuracy and less computational cost, was the variant in which the training set contained noise, but the validation set contained no noise. This option was also used in the present study.

To reduce the output data dimensionality, *autonomous determination* of the parameters [21] was used, where each parameter was determined separately from the rest. To reduce the input data dimensionality, we used selection of significant input features [22]. Thus, each of the ANN had a single output and a number of inputs, determined as the result of selection.

The problem was solved by perceptrons with 3 hidden layers containing 24, 16, 8 neurons. Sigmoidal transfer function was used in the hidden layers, and linear in the output layer. For each determined parameter, 5 ANNs were trained with various initializations of the weights; performance indexes of these 5 ANNs were averaged.

5   Results

Neural networks trained on data with noise of a certain level and statistics were applied to test data sets with various levels of noise of the same statistics.

Figure 1 shows the dependence of the solution quality (coefficient of multiple determination R^2) on the level of the noise in the test data set for networks trained on training sets with various noise levels, for the determined parameter y17 (upper layer of the grid), for various types and levels of the noise. Here one can see that the higher is the level of noise in the training dataset, the worse the network operates on noiseless data, but the slower it degrades with increasing noise level. For the other considered parameters, the nature of the dependency is completely similar.

Figure 2 presents the dependence of the solution quality (coefficient of multiple determination R^2) on the level of the noise in the training dataset for different noise levels in the test dataset, for the determined parameter y83 (third layer of the grid), for various types and levels of the noise. The character of the dependences shows that for each level of noise in the test data set, there is its optimal ANN, as a rule, the one trained with noise of the same level as the one contained in the considered test set.

As the effect of increasing noise resilience of neural network solutions of IP with adding noise to the training set was observed for two different types of IPs, we can conclude that this effect is connected with the fundamental properties of the perceptron, rather than with the properties of the data.

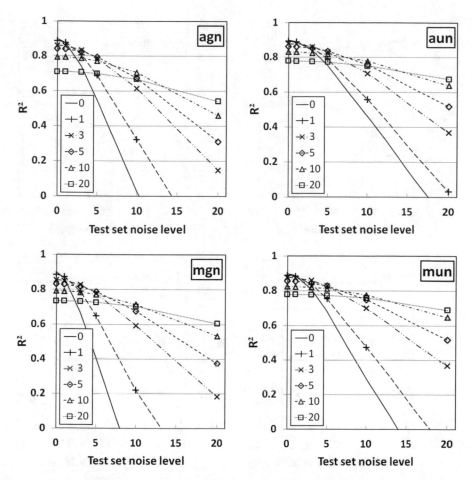

Fig. 1. The dependence of the solution quality (coefficient of multiple determination R^2) on the level of noise in the test data set for networks trained on training sets with various noise levels. Determined parameter y17. Additive Gaussian (agn), additive uniform (aun), multiplicative Gaussian (mgn), multiplicative uniform (mun) noise. Various curves correspond to the specified noise level in percent in the training set while training the networks.

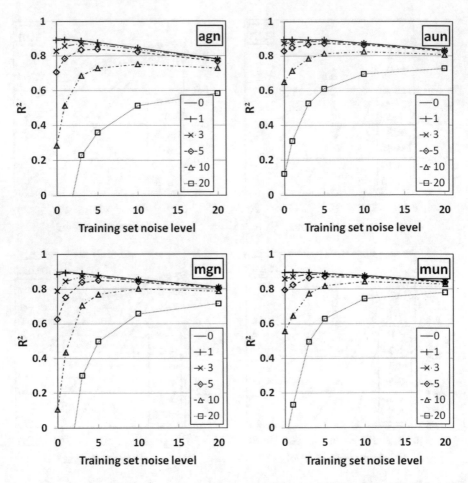

Fig. 2. The dependence of the solution quality (coefficient of multiple determination R^2) on the level of noise in the training data set for various noise levels in the test data set. Determined parameter y83. Additive Gaussian (agn), additive uniform (aun), multiplicative Gaussian (mgn), multiplicative uniform (mun) noise. Various curves correspond to the specified noise level in percent in the test dataset.

6 Conclusion

In this study, the existence of the effect of increasing noise resilience of a neural network solution of the inverse problem to noise in the input data with adding noise to the training set when training neural networks, has been confirmed on the data of IP of magnetotelluric sounding. The effect was observed for different distributions and at various intensities of the noise.

As the results of the study, we can draw the following conclusions:

1. The higher is the level of noise in the training data set, the worse the network operates on data without noise, but the more slowly it degrades with increasing noise level in the test data set.
2. Maximum noise resilience (maximum quality of the solution) is generally achieved when the level of the noise in the training data set coincides with the level of noise during network application (in the test dataset).
3. The observed effect is associated with the fundamental properties of multilayer perceptron neural networks and not with the properties of the data. So this method can be used solving other multi-parameter inverse problems.

Acknowledgement.   The authors would like to thank E.A. Obornev, I.E. Obornev, and M.I. Shimelevich for providing the data on which this study has been performed.

References

1. Zhdanov, M.: Inverse Theory and Applications in Geophysics, 2nd edn. Elsevier, Amsterdam (2015)
2. Yagola, A., Kochikov, I., Kuramshina, G.: Inverse Problems of Vibrational Spectroscopy. De Gruyter, Berlin (1999)
3. Mohammad-Djafari, A. (ed.): Inverse Problems in Vision and 3D Tomography. Wiley, New York (2010)
4. Spichak, V.V. (ed.): Electromagnetic Sounding of the Earth's Interior. Methods in Geochemistry and Geophysics, vol. 40. Elsevier, Amsterdam (2006)
5. Zhdanov, M.S.: Geophysical Electromagnetic Theory and Methods. Methods in Geochemistry and Geophysics, vol. 43. Elsevier, Amsterdam (2009)
6. Spichak, V., Popova, I.: Artificial neural network inversion of magnetotelluric data in terms of three-dimensional earth macroparameters. Geophys. J. Int. **142**(1), 15–26 (2000)
7. Li, M., Verma, B., Fan, X., Tickle, K.: RBF neural networks for solving the inverse problem of backscattering spectra. Neural Comput. Appl. **17**(4), 391–397 (2008)
8. Yang, H., Xu, M.: Solving inverse bimodular problems via artificial neural network. Inverse Probl. Sci. Eng. **17**(8), 999–1017 (2009)
9. Holmstrom, L., Koistinen, P.: Using additive noise in back-propagation training. Proc. IEEE Trans. Neural Netw. **3**(1), 24–38 (1992)
10. Matsuoka, K.: Noise injection into inputs in back-propagation learning. Proc. IEEE Trans. Syst. Man Cybern. **22**(3), 436–440 (1992)
11. An, G.: The effects of adding noise during backpropagation training on a generalization performance. Neural Comput. **8**(3), 643–674 (1996)
12. Zur, R.M., Jiang, Y., Pesce, L.L., Drukker, K.: Noise injection for training artificial neural networks: a comparison with weight decay and early stopping. Med. Phys. **36**(10), 4810–4818 (2009)
13. Piotrowski, A.P., Napiorkowski, J.J.: A comparison of methods to avoid overfitting in neural networks training in the case of catchment runoff modeling. J. Hydrol. **476**, 97–111 (2013)
14. Bishop, C.M.: Training with noise is equivalent to Tikhonov regularization. Neural Comput. **7**(1), 108–116 (1995)
15. Wang, C., Principe, J.C.: Training neural networks with additive noise in the desired signal. Proc. IEEE Trans. Neural Netw. **10**(6), 1511–1517 (1999)

16. Yin, S., Liu, C., Zhang, Z., Lin, Y., Wang, D., Tejedor, J., Zheng, T.F., Li, Y.: Noisy training for deep neural networks in speech recognition. Proc. EURASIP J. Audio Speech Music Process. **2015**(2), 1–14 (2015)
17. Fadeev, V.V., Dolenko, S.A., Dolenko, T.A., Uvenkov, Ya.V., Filippova, E.M., Chubarov, V.V.: Laser diagnostics of complicated organic compounds and complexes by saturation fluorimetry. Quantum Electron. **27**(6), 556–559 (1997)
18. Dolenko, S.A., Dolenko, T.A., Kozyreva, O.V., Persiantsev, I.G., Fadeev, V.V., Filippova, E.M.: Solution of inverse problem in nonlinear laser fluorimetry of organic compounds with the use of artificial neural networks. Pattern Recognit. Image Anal. **9**(3), 510–515 (1999)
19. Gerdova, I.V., Dolenko, S.A., Dolenko, T.A., Persiantsev, I.G., Fadeev, V.V., Churina, I.V.: New opportunity solutions to inverse problems in laser spectroscopy involving artificial neural networks. Izvestiya Akademii Nauk. Ser. Fizicheskaya **66**(8), 1116–1125 (2002)
20. Isaev, I.V., Dolenko, S.A.: Training with noise as a method to increase noise resilience of neural network solution of inverse problems. Opt. Mem. Neural Netw. **25**(3), 142–148 (2016)
21. Dolenko, S., Isaev, I., Obornev, E., Persiantsev, I., Shimelevich, M.: Study of influence of parameter grouping on the error of neural network solution of the inverse problem of electrical prospecting. Commun. Comput. Inf. Sci. **383**, 81–90 (2013)
22. Dolenko, S., Guzhva, A., Obornev, E., Persiantsev, I., Shimelevich, M.: Comparison of adaptive algorithms for significant feature selection in neural network based solution of the inverse problem of electrical prospecting. In: Alippi, C. (ed.) ICANN 2009, Part II. Springer-Verlag, Heidelberg (2009)

Multi-Layer Solution of Heat Equation

Tatiana Lazovskaya[1]([⊠]), Dmitry Tarkhov[2], and Alexander Vasilyev[2]

[1] Computing Center of FEB RAS, Khabarovsk, Russia
tatianala@list.ru
[2] Peter the Great Saint-Petersburg Politechnical University, Saint Petersburg, Russia

Abstract. A new approach to the construction of multilayer neural network approximate solutions for evolutionary partial differential equations is considered. The approach is based on the application of the recurrence relations of the Euler, Runge-Kutta, etc. methods to variable length intervals. The resulting neural-like structure can be considered as a generalization of a feedforward multilayer network or a recurrent Hopfield network. This analogy makes it possible to apply known methods to the refinement of the obtained solution, for example, the backpropagation algorithm. Earlier, a similar approach has been successfully used by the authors in the case of ordinary differential equations. Computational experiments are performed on one test problem for the one-dimensional (in terms of spatial variables) heat equation. Explicit formulas are derived for the dependence of the resulting neural network output on the number of layers. It was found that the error tends to zero with an increasing number of layers, even without the use of the network learning.

Keywords: Partial differential equation · Approximate solution · Multilayer solution · Neural network · One-dimensional heat equation

1 Introduction

Let us consider an evolution equation of the form

$$\frac{\partial}{\partial t} u(x,t) = F(u(x,t), x, t), \qquad (1)$$

where $u(x,t)$ is a sufficiently smooth function with respect to the variables $(x,t) \in R^p \times R^+$ and F is some linear mapping, for example, a differential operator. It is assumed that the required solution $u(x,t)$ satisfies the initial condition $u(x,0) = \varphi(x)$. Other additional conditions, such as belonging to a certain functional space, boundary conditions, etc., can be omitted in the context we are considering. Discussion of these conditions makes sense in solving a particular problem. We construct approximate solutions of problem (1) as a function of time (with conditionally fixed x) for the interval $[0,t]$ with variable right end. For this, we use the recursive scheme of the implicit Euler method of the form

$$u_{k+1}(x) = u_k(x) + hF(u_{k+1}(x), x, t_{k+1}), \qquad (2)$$

© Springer International Publishing AG 2018
B. Kryzhanovsky et al. (eds.), *Advances in Neural Computation, Machine Learning, and Cognitive Research*, Studies in Computational Intelligence 736,
DOI 10.1007/978-3-319-66604-4_3

with a constant step $h = t/n$, where n is called the number of layers. You can also use other methods, such as the Runge-Kutta method. As an approximation to the solution, we propose to use $u_n(x, t)$. We denote by $u_k(x)$ the value of the solution on the layer with number k (at $t_k = kt/n$) and obtain a series of equations for approximations $u_k(x) = u(x, t_k)$

$$\frac{1}{h}u_{k+1}(x) - F(u_{k+1}(x), x, t_k) = \frac{1}{h}u_k(x), \ k = 0, \dots, n-1;$$
$$u_0(x) = u(x, 0) = \varphi(x).$$

To solve these equations, one can apply any suitable method, for example, the Newton method. As a result, we obtain a recurrence relation of the form $u_{k+1}(x, t) = G(u_k(x, t), x, t_{k+1})$ that can be interpreted as a procedure for the transition from layer to layer of a neural-like structure. In this case, the similarity with neural networks increases if the operator can be represented in integral form $G(u, x, t) = \int K(u, x, t, y, s)dyds$. Thus, we obtain a neural network with a continuum of neurons, similar to that considered in [17] for other problems. The final neural structure can be considered as the generalization of multilayered backpropagation networks or recurrence Hopfield networks [16]. Such an analogy allows us to apply to the refinement of the resulting solution (an operator G or a kernel K) known methods, for example, the back propagation algorithm.

In the case when the mapping F is representable in the form $(u(x, t), x, t) = Lu(x, t) + f(x, t)$, where L is some linear operator, for example, a linear differential operator with respect to the spatial variable x, we can proceed to the next series of approximations

$$(\frac{1}{h}I - L)u_{k+1}(x) = \frac{1}{h}u_k(x) + f_{k+1}(x), \ k = 0, \dots, n-1. \tag{3}$$

Here, I is an identity mapping and $f_{k+1}(x) = f(x, t_{k+1})$ is some known function. The value $u_0(x) = \varphi(x)$ is still considered as a given. If $\lambda = 1/h$ does not belong to the spectrum of the operator L then the operator resolvent $R = R(L) = (L - \frac{1}{h}I)^{-1}$ exist and approximations are given by formulas

$$u_1(x) = -\frac{1}{h}R\varphi(x) - Rf_1(x),$$
$$u_{k+1}(x) = -\frac{1}{h}Ru_k(x) - Rf_{k+1}(x), k = 1, \dots, n-1.$$

In the case where the resolvent can be represented in integral form, the result can be interpreted as a neural network with a continuum of neurons and apply the corresponding learning algorithms to refine the solution.

We will be interested in the case of an ordinary differential operator L on the semiaxis, that is,

$$\frac{\partial}{\partial t}u(x, t) = Lu(x, t), \tag{4}$$

where $(x, t) \in R^+ \times R^+$, L is a linear differential operator with respect to the spatial variable x. As a boundary condition, we consider the boundedness of the solution at infinity. A series of recurrence relations takes the form

$$Lu_{k+1}(x) - \frac{1}{h}u_{k+1}(x) = -\frac{1}{h}u_k(x). \tag{5}$$

The problem of constructing the resolvent for this case solved comprehensively.

This approach is illustrated using a simple model problem that has a known analytical solution.

2 Model Problem

As an example, we consider the classical heat equation [1] in the dimensionless form

$$\frac{\partial u(x,t)}{\partial t} = \frac{\partial^2 u(x,t)}{\partial x^2}, \tag{6}$$

where $x \in R$ and $t \in R^+$. We solve the initial-boundary value problem with a piecewise-given initial condition

$$u(x,0) = \theta(x) = \begin{cases} 1, & x \geq 0, \\ 0, & x < 0. \end{cases} \tag{7}$$

Function $\theta(x)$ is one of the typical activation functions widely used when the neural networks applying. The linearity of the problem under consideration makes it possible to reduce to this case any problem with an initial condition for which the function $u(x,0)$ is representable as a neural network with one hidden layer whose neurons have the activation function $\theta(x)$.

The solution will be sought among functions that are bounded at infinity. Using the methods of [2], we will construct an approximate solution based on the implicit Euler method [3] on the time interval with the variable right end $[0, t]$ (taking into account the initial condition (7)). Step $h = t/N$ varies depending on the selected number of iterations N. Applying the implicit Euler method with respect to the time variable t we obtain a second-order linear equation with respect to the variable x

$$u''_{n+1}(x) - \frac{1}{h}u_{n+1}(x) = -\frac{1}{h}u_n(x), \ 0 \leq n \leq N. (4) \tag{8}$$

Direct calculations showed that $u_1(x) = \begin{cases} 1 - 1/2\exp(-x/\sqrt{h}), & x \geq 0; \\ 1/2\exp(x/\sqrt{h}), & x \leq 0. \end{cases}$

Let us derive recurrent formulas for solving an equation of the type (8). For simplicity, we first make the change of variables. We denote $z := x/\sqrt{h}$, then the Eq. (8) has the form

$$u''_{n+}(z) - u_{n+1}(z) = -u_n(z), \tag{9}$$

where $u_1(z) = \begin{cases} 1 - 1/2\exp(-z), & z \geq 0; \\ 1/2\exp(z), & z \leq 0. \end{cases}$ Taking into account the properties of

a second-order linear equation of the form (9), in general form, it is necessary to write out the solution for the following problem

$$y''(z) - y(z) = A + P_m^+(z)\exp(z) + P_m^-(z)\exp(-z). \tag{10}$$

Here, $P_m^+(z) = \sum_{i=0}^m p_i^+ z^i$, $P_m^-(z) = \sum_{i=0}^m p_i^- z^i$ are polynomials of degree m.

Let us find a particular solution for the term $P_m^+(z)\exp(z)$. Since the number $\lambda = 1$ is a root of $k = 1$ multiplicity of the characteristic polynomial $\chi(\lambda) = \lambda^2 - 1$ for Eq. (10), we seek a particular solution by the method of undetermined coefficients in the form $y^*(z) = zQ_m^+(z)\exp(z) = \exp(z)\sum_{i=0}^m q_i^+ z^{i+1}$.

We substitute the representation for the solution $y^*(z)$ into Eq. (10) leaving only the term $P_m^+(z)\exp(z)$ and arrive at equality

$$\left(\exp(z)\sum_{i=0}^m q_i^+ z^{i+1}\right)'' - \exp(z)\sum_{i=0}^m q_i^+ z^{i+1} = \exp(z)\sum_{i=0}^m p_i^+ z^i. \tag{11}$$

After simplifications, we obtain the relations for the required coefficients q_i^+ in terms of the given p_i^+. From these relations it follows that

$$q_m^+ = \frac{p_m^+}{2(m+1)}; \tag{12}$$

$$q_i^+ = \frac{p_i^+}{2(i+1)} - q_{i+1}^+ \frac{i+2}{2}, \quad i = 0, \ldots, m-1. \tag{13}$$

For the term $P_m^-(z)\exp(-z)$ on the right-hand side of the equation, we similarly have a representation for solving

$$y^*(z) = zQ_m^-(z)\exp(-z) = \exp(-z)\sum_{i=0}^m q_i^- z^{i+1}. \tag{14}$$

The method of undetermined coefficients leads to relations, analogous to formulas (12)–(13).

3   Calculations

Using formula (13), we can easily calculate the approximate solution from relations (8) for any values N of the number of iterations. Note that the exact solution of the problem (6)–(7) has the form

$$u(x,t) = \frac{1}{2} - \frac{1}{\sqrt{\pi}}\int_0^z \exp(-t^2)dt. \tag{15}$$

General form of solution obtained after the use of N iterations looks like

$$u_N(x,t) = \begin{cases} 1 - P_{N-1}^+\left(\frac{x}{\sqrt{t/N}}\right)\exp\left(-\frac{x}{\sqrt{t/N}}\right), & x \geq 0, \\ P_{N-1}^-\left(\frac{x}{\sqrt{t/N}}\right)\exp\left(\frac{x}{\sqrt{t/N}}\right), & x \leq 0. \end{cases} \tag{16}$$

In Fig. 1, we can clearly evaluate the order of approximation of the method. The graphs of errors for different values of the number of iterations and at different instants of time are presented. We note that the order of the maximum error of a particular solution for a fixed number of iterations does not depend on the chosen time moment t.

Note that there is a tendency $|u_N(x,t) - u(x,t)| < 10^{-(1+[N/10])}$.

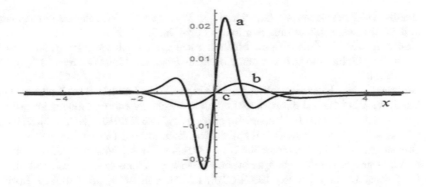

Fig. 1. Error graph of the approximate solution of problem (6)–(7) obtained by (8) in the case of $x \in [-5;5]$ and (a) $N = 3$; $t = 0,1$, (b) $N = 20$, $t = 0,5$, (c) $N = 1000$, $t = 20$.

4   Conclusions

It should be noted that the above formulas should be considered as a direct operation of the neural network with initially given weights. If the received accuracy is not enough, then you can apply the usual learning procedure (for example, the error back propagation method). In this case, the numerical parameters of formulas (10)–(13) are considered as weights and an appropriate error functional is minimized [5–15]. The construction of a solution of a differential equation in the form of an initially multi-layer functional model with the possibility of subsequent training is a breakthrough method [16].

The proposed neural network approach seems promising. It is known [4] that any function from $L^2(R)$ can be approximated with a given accuracy by means of the corresponding weighted sum of shifts of functions of the form (7). Moreover, instead of functions of the form (7) under initial conditions, one can consider sigmoidal or other functions that are used as the basis function of the neural network with a single hidden layer in the approximation of a given function. In this case, for such an approximation of the initial condition, the general scheme of the solution of problem (1) described in the article will be preserved. External influences on each of the layers can also be taken into account and calculated using Duhamel's integrals.

References

1. Tikhonov, A.N., Samarskiy, A.A.: Uravneniya Matematicheskoy Fiziki. Izdatel'stvo MGU, Moscow (1999). (in russian)
2. Lazovskaya, T., Tarkhov, D.: Multilayer neural network models based on grid methods. IOP Conf. Ser.: Mater. Sci. Eng. **158**(1) (2016). doi:10.1088/1757-899X/158/1/012061
3. Samarskiy, A.A.: Vvedeniye v chislennyye metody. Lan', SPb (2005). (in russian)

4. Hardle, W., Kerkyacharian, G., Picard, D., Tsybakov, A.: Wavelets, Approximation and Statistical Applications. Seminaire Paris-Berlin (1997)
5. Vasilyev, A.N., Tarkhov, D.A.: Mathematical models of complex systems on the basis of artificial neural networks. Nonlinear Phenom. Complex Syst. $17(3)$, 327–335 (2014)
6. Budkina, E.M., Kuznetsov, E.B., Lazovskaya, T.V., Leonov, S.S., Tarkhov, D.A., Vasilyev, A.N.: Neural network technique in boundary value problems for ordinary differential equations. In: Cheng, L., et al. (eds.) ISNN 2016. LNCS, vol. 9719, pp. 277–283. Springer International Publishing, Switzerland (2016)
7. Gorbachenko, V.I., Lazovskaya, T.V., Tarkhov, D.A., Vasilyev, A.N., Zhukov, M.V.: Neural network technique in some inverse problems of mathematical physics. In: Cheng, L., et al. (eds.) ISNN 2016. LNCS, vol. 9719, pp. 310–316. Springer International Publishing, Switzerland (2016)
8. Shemyakina, T.A., Tarkhov, D.A., Vasilyev, A.N.: Neural network technique for processes modeling in porous catalyst and chemical reactor. In: Cheng, L., et al. (eds.) ISNN 2016. LNCS, vol. 9719, pp. 547–554. Springer International Publishing, Switzerland (2016)
9. Kaverzneva, T., Lazovskaya, T., Tarkhov, D., Vasilyev, A.: Neural network modeling of air pollution in tunnels according to indirect measurements. J. Phys.: Conf. Ser. 772 (2016). doi:10.1088/1742-6596/772/1/012035
10. Lazovskaya, T.V., Tarkhov, D.A., Vasilyev, A.N.: Parametric neural network modeling in engineering. Recent Patents Eng. $11(1)$, 10–15 (2017)
11. Bolgov, I., Kaverzneva, T., Kolesova, S., Lazovskaya, T., Stolyarov, O., Tarkhov, D.: Neural network model of rupture conditions for elastic material sample based on measurements at static loading under different strain rates. J. Phys.: Conf. Ser. 772 (2016). doi:10.1088/1742-6596/772/1/012032
12. Filkin, V., Kaverzneva, T., Lazovskaya, T., Lukinskiy, E., Petrov, A., Stolyarov, O., Tarkhov, D.: Neural network modeling of conditions of destruction of wood plank based on measurements. J. Phys.: Conf. Ser. 772 (2016). doi:10.1088/1742-6596/772/1/012041
13. Vasilyev, A., Tarkhov, D., Bolgov, I., Kaverzneva, T., Kolesova, S., Lazovskaya, T., Lukinskiy, E., Petrov, A., Filkin, V.: Multilayer neural network models based on experimental data for processes of sample deformation and destruction. In: Selected Papers of Convergent 2016, Moscow, 25–26 November 2016, pp. 6–14 (2016)
14. Tarkhov, D., Shershneva, E.: Approximate analytical solutions of Mathieus equations based on classical numerical methods. In: Selected Papers of SITITO 2016, Moscow, 25–26 November 2016, pp. 356–362 (2016)
15. Vasilyev, A., Tarkhov, D., Shemyakina, T.: Approximate analytical solutions of ordinary differential equations. In: Selected Papers of SITITO 2016, Moscow, 25–26 November 2016, pp. 393–400 (2016)
16. Neha, Y., Anupam, Y., Manoj, K.: An Introduction to Neural Network Methods for Differential Equations. Springer Briefs in Applied Sciences and Technology Computational Intelligence (2015)
17. Galushkin, A.I.: Neyronnyye seti: osnovy teorii. Goryachaya liniya - telekom, Moscow (2010). (in Russian)

Implementation of a Gate Neural Network Based on Combinatorial Logic Elements

Taras Mikhailyuk$^{(\boxtimes)}$ and Sergey Zhernakov

Ufa State Aviation Technical University, Ufa, Russia
`realotoim@mail.ru`

Abstract. Generally, math models which use the "continuous mathematics" are dominant in the construction of modern digital devices, while the discrete basis remain without much attention. However, when solving the problem of constructing effective computing devices it is impossible to ignore the compatibility level of the mathematical apparatus and the computer platform used for its implementation. In the field of artificial intelligence, this problem becomes urgent during the development of specialized computers based on the neural network paradigm. In this paper, the disadvantages of the application of existing approaches to the construction of a neural network basis are analyzed. A new method for constructing a neural-like architecture based on discrete trainable structures is proposed to improve the compatibility of artificial neural network models in the digital basis of programmable logic chips and general-purpose processors. A model of a gate neural network using a mathematical apparatus of Boolean algebra is developed. Unlike formal models of neural networks, proposed network operates with the concepts of discrete mathematics. Formal representations of the gate network are derived. The learning algorithm is offered.

Keywords: Boolean algebra · Boolean neural network · Combinatorial logic · Delta rule · Gate neural network · Logical network · Widrow Hoff rule

1 Introduction

Quite often in practice, there are problems associated with the compatibility of the functional and hardware-software parts of the device. These problems are very complex and require an integrated approach. Their solution leads to a change in qualitative and quantitative characteristics according to specified requirements.

Artificial intelligence algorithms require complex use of hardware and software. Due to specific nature of the research, such basic indicators as productivity, diminutiveness and low economic costs associated with the production and maintenance of the devices being developed remain unchanged. The approach based on modeling of artificial neural networks is versatile and flexible, but has limitations related to the field of their application. Among the disadvantages inherent to the computer of von Neumann architecture, we can distinguish the following:

- virtualization of calculators, architecture, physical processes;
- the dependence of the processing time on the size of the program;

© Springer International Publishing AG 2018
B. Kryzhanovsky et al. (eds.), *Advances in Neural Computation, Machine Learning, and Cognitive Research*, Studies in Computational Intelligence 736,
DOI 10.1007/978-3-319-66604-4_4

- unjustified growth of hardware costs when increasing productivity;
- low energy efficiency, etc.

At present, there is an increasing number of specialized intellectual architectures aimed at overcoming the described drawbacks [1–8]. Such devices have wide application range and are compatible with the environment of the computer system, but they also have some disadvantages. Generally, math models which use the "continuous mathematics" are dominant in the construction of modern digital devices, while the discrete basis remain without much attention. However, solving the problem of constructing effective computing devices it is impossible to ignore the compatibility level of the mathematical apparatus and the computer platform used for its implementation. In the field of artificial intelligence, this problem becomes urgent during the development of specialized computers based on the neural network paradigm.

Existing mathematical models of a neuron operate with continuous quantities, are realized on the basis of an analog elements, which leads to their poor compatibility with digital equipment. But at the same time, most neural networks use the principles of digital logic [2–4, 6–8]. And as the result, in promising computing devices being developed multi-level systems of models are implemented. These systems introduce certain disadvantages in the final implementation of the solution [9, 10].

In this paper, a method for constructing a neural-like architecture based on discrete trainable structures is proposed to improve the compatibility of artificial neural network models in the digital basis of programmable logic chips and general-purpose processors.

2 Model of the Gate Neural Network

The trainable gate network is representative of Boolean networks [5, 11–16] with the ability to specify the type of mapping of the vector of input signals to the output vector, using the learning algorithm. Such a network can be considered as an attempt to combine certain features of neural network technology and combinational logic gates to achieve a synergistic effect in the implementation of high-performance embedded systems.

We obtain a formalized representation of this type of network. It is known from dicrete mathematics that the full disjunctive normal form (FDNF) can be represented as follows:

$$f(x_1, \ldots, x_P) = \bigvee_{\substack{(\sigma_1, \ldots, \sigma_P) \\ f(\sigma_1, \ldots, \sigma_P) = 1}} x_1^{\sigma_1} \wedge, \ldots, \wedge x_P^{\sigma_P}, \tag{1}$$

while the disjunction of all sets has the form:

$$y = f(\sigma_1, \ldots, \sigma_P) = 1 \tag{2}$$

Rule (2) can be reformulated as a disjunction over all full product terms (FPT) of P variables:

$$\overset{2^P}{\underset{n=1}{\vee}} \psi_n(\mathbf{x}) = 1 \tag{3}$$

Then the minimal term can be written in the following way:

$$\psi_n(\mathbf{x}) = \overset{P}{\underset{p=1}{\wedge}} x_p^{M_p(n-1)} \tag{4}$$

Next, we define the function Mp (α):

$$M_p(\alpha) = \begin{cases} 0, & \alpha \in [i \cdot T; \ (i+0,5) \cdot T), \\ 1, & \text{otherwise.} \end{cases} \tag{5}$$

where the period $T = 2^p$, $i = 0, 1, 2, \ldots, \frac{2^p}{T} - 1$.

The function (5) is square wave logical basis, similar to the Rademacher function [17]. Figure 1 shows the form of this function for $p \le 3$.

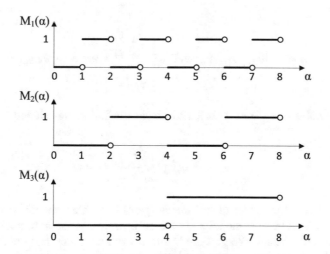

Fig. 1. View of the square wave function for $p \le 3$

The square wave function masks each variable included in Eq. (4) with the goal of specifying all FPTs. Next, we represent the FPT (3) in vector form:

$$\psi = [\psi_1(\mathbf{x}), \ \psi_2(\mathbf{x}), \ \ldots, \ \psi_N(\mathbf{x})], \tag{6}$$

where **x**—the column vector of input signals:

$$\mathbf{x} = [x_1, \ x_2, \ \ldots, \ x_P]^T. \tag{7}$$

Next, we weigh functions of input signals in vector form, which is known from the theory of neural networks [1, 18, 19]:

$$\mathbf{w}^T \wedge \boldsymbol{\psi} = \mathbf{y}, \tag{8}$$

where **w**—the column vector (9), and **y**—the column vector (10):

$$\mathbf{w} = [w_1, \ w_2, \ \ldots, \ w_N]^T, \tag{9}$$

$$\mathbf{y} = [y_1, \ y_2, \ \ldots, \ y_S]^T. \tag{10}$$

The matrix Eq. (8) has a similar form with the equation describing the formal neuron, radial basis function network [18, 19] and also the sigma-pi network [20], but in this case the multiplication operation is replaced by the conjunction operation, since the matrices have a binary form.

For a network containing one element in the output layer, we get the following expression:

$$\mathbf{w}^T \wedge \boldsymbol{\psi} = [w_1, \ w_2, \ \ldots, \ w_N] \wedge \begin{bmatrix} \boldsymbol{\psi}_1(\mathbf{x}) \\ \boldsymbol{\psi}_2(\mathbf{x}) \\ \ldots \\ \boldsymbol{\psi}_N(\mathbf{x}) \end{bmatrix} = \bigvee_{n=1}^{N} w_n \wedge \boldsymbol{\psi}_n(\mathbf{x}). \tag{11}$$

Next, we substitute (4) into (11), and obtain the following relation in the general form:

$$y = \bigvee_{n=1}^{N} w_n \wedge \bigwedge_{p=1}^{P} x_p^{M_p(n-1)} \tag{12}$$

The Eq. (12) is the model of a Boolean (gate) trainable network. It follows from expression (12) that in such model there are no operators inherent to neural networks, since they are bit-oriented. Weights are Boolean variables there, and not real numbers. This model describes a two-layer network in which the first layer is represented by a set of N constituent units (4), besides this layer does not require training. The output layer is represented by one disjunctive element, which summarizes the minterms, enabled by means of weight coefficients.

A similar dependence can be obtained for a network with several elements in the output layer:

$$\mathbf{w}^T \wedge \boldsymbol{\psi} = \begin{bmatrix} w_{11}, & w_{12}, & \ldots, & w_{1N} \\ w_{21}, & w_{22}, & \ldots, & w_{2N} \\ & & \ldots & \\ w_{S1}, & w_{S2}, & \ldots, & w_{SN} \end{bmatrix} \wedge \begin{bmatrix} \boldsymbol{\psi}_1(\mathbf{x}) \\ \boldsymbol{\psi}_2(\mathbf{x}) \\ \ldots \\ \boldsymbol{\psi}_N(\mathbf{x}) \end{bmatrix} = \begin{bmatrix} \overset{N}{\underset{n=1}{\vee}} w_{1n} \wedge \boldsymbol{\psi}_n(\mathbf{x}) \\ \overset{N}{\underset{n=1}{\vee}} w_{2n} \wedge \boldsymbol{\psi}_n(\mathbf{x}) \\ \ldots \\ \overset{N}{\underset{n=1}{\vee}} w_{sn} \wedge \boldsymbol{\psi}_n(\mathbf{x}) \end{bmatrix} = \begin{bmatrix} y_1 \\ y_2 \\ \ldots \\ y_S \end{bmatrix}.$$

$$(13)$$

Then the Eq. (6) for each output can be written in a general form:

$$y_s = \overset{N}{\underset{n=1}{\vee}} w_{sn} \wedge \overset{P}{\underset{p=1}{\wedge}} x_p^{M_p(n-1)} \qquad (14)$$

The analysis of dependences (13) and (14) shows that it is possible to synthesize on their basis an arbitrary combination device with P inputs and S outputs, which has two levels of gates and has an increased speed in hardware implementation. These formulas represent a trainable logical basis. Figure 2 shows a graph of the network.

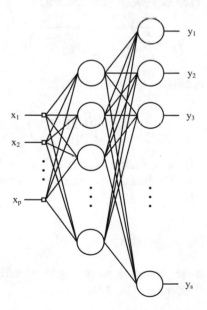

Fig. 2. Trainable gate neural network

It is known that the maximum number of combinations of P variables is equal to 2^P, and the number of functions is 2^{2^P}. It follows that the number of neurons of the first layer is not more than 2^P:

$$N \leq 2^P. \tag{15}$$

In turn, the number of neurons in the output layer is less than 2^{2^P}:

$$S \leq 2^{2^P}. \tag{16}$$

Thus, the maximum sum from (15) and (16) describes the largest network without repeating elements. However, duplication of elements can be aimed to increasing the reliability of the network.

It is not difficult to show that the obtained model can be realized in the form of a full conjunctive normal form (FCNF). On the basis of de Morgan's laws for several variables [21], we can show:

$$\overset{N}{\underset{n=1}{\vee}} a_n = \overline{\overset{N}{\underset{n=1}{\wedge}} \bar{a}_n}. \tag{17}$$

Applying the rule (17) to expression (12) we obtain:

$$y_s = \overline{\overset{N}{\underset{n=1}{\wedge}} \left(\bar{w}_{sn} \vee \overset{P}{\underset{p=1}{\vee}} x_p^{\bar{M}_p(n-1)} \right)}. \tag{18}$$

Next, replacing the variables, we get the FCNF:

$$\lambda_s = \overset{N}{\underset{n=1}{\wedge}} \left(m_{sn} \vee \overset{P}{\underset{p=1}{\vee}} x_p^{W_p(n-1)} \right). \tag{19}$$

Equations (12) and (19) are equivalent in essence like the FCNF and the FDNF are equivalent. It is seen from (19) that the weighing is performed by the disjunction operation, in contrast to (12).

3   Network Learning Algorithm

The learning algorithm of the perceptron according to the Widrow-Hoff rule is known from the theory of neural networks, [18, 19]:

$$w_{sn}(t+1) = w_{sn}(t) + \Delta w_{sn}(t), \tag{20}$$

$$\Delta w_{sn}(t) = x_n(t) \cdot (d_s - y_s(t)), \tag{21}$$

On the basis of (20) and (21), it is easy to see the following:

- weight w_{sn} can increase or decrease depending on the sign of the increment of weight Δw_{sn};
- weight change occurs when the output signal y_s deviates from the reference d_s only for the input x_n which causes this influence.

Using these statements, we can show the training algorithm for a binary network. We convert these formulas into a system of residual classes. It is known that additive operations and multiplication will look like the following [22]:

$$(a \pm b) \bmod c = ((a \bmod c) \pm (b \bmod c)) \bmod c, \tag{22}$$

$$(a \cdot b) \bmod c = ((a \bmod c) \cdot (b \bmod c)) \bmod c. \tag{23}$$

We describe (20) and (21), using (22) and (23). Then the Widrow-Hoff rules will take the form which is typical for operations performed by digital devices:

$$\begin{aligned} w_{sn}(t+1) \bmod q &= (w_{sn}(t) + \Delta w_{sn}(t)) \bmod q \\ &= (w_{sn}(t) \bmod q + \Delta w_{sn}(t) \bmod q) \bmod q, \end{aligned} \tag{24}$$

$$\begin{aligned} \Delta w_{sn}(t) \bmod q &= (x_n(t) \cdot (d_s - y_s(t))) \bmod q \\ &= ((x_n(t) \bmod q) \cdot ((d_s) \bmod q - y_s(t) \bmod q)) \bmod q, \end{aligned} \tag{25}$$

where q is a positive integer.

It is required that all variables (24) and (25) could accept only two states, or that the modulo is equal 2. Considering that additive operations can be replaced by the exclusive-OR operation and multiplication—by conjunctions, the Widrow-Hoff rule will be written in the following form:

$$w_{sn}(t+1) = w_{sn}(t) \oplus x_n(t) \wedge (d_s \oplus y_s(t)). \tag{26}$$

We apply rule (26) to the received network model (12). Taking into account the influence of minterms (4) on the learning element, we obtain the learning rule for the Boolean network:

$$w_{sn}(t+1) = w_{sn}(t) \oplus (d_s \oplus y_s(t)) \wedge \bigwedge_{p=1}^{P} (x_p(t))^{M_p(n-1)}. \tag{27}$$

4 Analysis of the Results

On the basis of the dependence (12), the following features of the model can be noted:

- the model is a network;
- first and second layer have specialization;
- signals can be either excitatory or inhibitory;
- the type of generalization is different for FDNF and FCNF networks;

- there is no influence of minterms (maxterms) on each other.

Unlike formal models of neural networks, the Boolean network operates with the concepts of discrete mathematics. From the point of view of an intelligent approach, only binary input signals processing may seem insufficient when working with higher-order sets, but the feature of the obtained formulas (12), (19) is in the possibility of applying them as a logical basis controlled by weight coefficients. It is known that on the basis of a Boolean basis arbitrary combinational devices are constructed. Furthermore, with the actual implementation of the trainable gate network, it is characterized by greater performance and reliability associated with the fixed depth of the gates and the simplicity of the individual handlers. For solving more complicated tasks it is possible to use the series of gate networks. In this case, the topology of the device is more homogeneous, which leads to the interchangeability of its individual elements.

The developed network can be considered as a basis for constructing feedforward neural networks with a flexible topology that can be adapted to a specific task, up to the level of logical elements.

The proposed approach has the following advantages:

1. Greater homogeneity of the topology of the device, in contrast to the formal neuron, which contains adders, multipliers, activation functions.
2. Increase of the applied component on the hardware level to solve specific problems.
3. Reduction of the occupied area of the crystal, which is required for the hardware implementation of the network.
4. Parallelizing of the processing and learning of the network at the level of logical elements.
5. Flexible learning architecture of a formal neuron.

5   Conclusion

The work in the field of creating discrete learning networks is aimed to solve the problems of optimizing hardware and software costs in the construction of neural networks and digital equipment in general. The trainable gate network is not intended to replace a feedforward neural network, but it can be considered as a basis for constructing any digital network. The possibilities of gate networks are quite various. They can find the application for the creation of associative memory devices, cryptography, high performance combinational devices, solvers of Boolean functions and in other applications.

References

1. Aljautdinov, M.A., Galushkin, A.I., Kazancev, P.A., Ostapenko, G.P.: Neurocomputers: from software to hardware implementation, p. 152. Gorjachaja linija - Telekom, Moscow (2008). (in Russian)
2. Mezenceva, O.S., Mezencev, D.V., Lagunov, N.A., Savchenko, N.S.: Implementations of non-standard models of neuron using Neuromatrix. Izvestija JuFU. Tehnicheskie nauki **131** (6), 178–182 (2012). (in Russian)
3. Adetiba, E., Ibikunle, F.A., Daramola, S.A., Olajide, A.T.: Implementation of efficient multilayer perceptron ANN neurons on field programmable gate array chip. Int. J. Eng. Technol. **14**(1), 151–159 (2014)
4. Manchev, O., Donchev, B., Pavlitov, K.: FPGA implementation of artificial neurons. Electronics: An Open Access Journal, Sozopol, Bulgaria, 22–24 September (2004). https://www.researchgate.net/publication/251757109_FPGA_IMPLEMENTATION_OF_ARTIFICIAL_NEURONS. Accessed 28 Jan 2017
5. Kohut R., Steinbach B.: The Structure of Boolean Neuron for the Optimal Mapping to FPGAs. http://www.informatik.tu-freiberg.de/prof2/publikationen/CADSM2005_BN_FPGA.pdf. Accessed 1 Feb 2017
6. Korani, R., Hajera, H., Imthiazunnisa, B., Chandra Sekhar, R.: FPGA modelling of neuron for future artificial intelligence applications. Int. J. Adv. Res. Comput. Commun. Eng. **2**(12), 4763–4768 (2013)
7. Omondi, A. R., Rajapakse, J. C.: FPGA Implementations of Neural Networks. Springer (2006). http://lab.fs.uni-lj.si/lasin/wp/IMIT_files/neural/doc/Omondi2006.pdf. Accessed 28 Jan 2017
8. Gribachev, V.: Element base of hardware implementations of neural networks (in Russian). http://kit-e.ru/articles/elcomp/2006_8_100.php. Accessed 30 June 2016
9. Mikhailyuk, T. E., Zhernakov, S. V.: Increasing efficiency of using FPGA resources for implementation neural networks. In: Nejrokomp'jutery: razrabotka, primenenie, vol. 11, pp. 30–39 (2016). (in Russian)
10. Mikhailyuk, T.E., Zhernakov, S.V.: On an approach to the selection of the optimal FPGA architecture in neural network logical basis. Informacionnye tehnologii **23**(3), 233–240 (2017). (in Russian)
11. Kohut, R., Steinbach, B.: Decomposition of Boolean Function Sets for Boolean Neural Networks. https://www.researchgate.net/publication/228865096_Decomposition_of_Boolean_Function_Sets_for_Boolean_Neural_Networks. Accessed 1 Feb 2017
12. Anthony, M.: Boolean Functions and Artificial Neural Networks. http://www.cdam.lse.ac.uk/Reports/Files/cdam-2003-01.pdf. Accessed 29 Jan 2017
13. Kohut, R., Steinbach, B.: Boolean neural networks. WSEAS Trans. Syst. **3**(2), 420–425 (2004)
14. Steinbach, B., Kohut, R.: Neural Networks – A Model of Boolean Functions. https://www.researchgate.net/publication/246931125_Neural_Networks_-_A_Model_of_Boolean_Functions. Accessed 1 Feb 2017
15. Vinay, D.: Mapping boolean functions with neural networks having binary weights and zero thresholds. IEEE Trans. Neural Netw. **12**(3), 639–642 (2001)
16. Zhang, C., Yang, J., Wu, W.: Binary higher order neural networks for realizing boolean functions. IEEE Trans. Neural Netw. **22**(5), 701–713 (2011)
17. Rademacher, H.: Einige Sätze über Reihen von allgemeinen Orthogonalfunktionen. Math. Ann. **87**(1–2), 112–138 (1922)

18. Hajkin, S.: Neural networks: a comprehensive foundation. Vil'jams, Moscow (2008). (in Russian)
19. Osovskij, S.: Neural networks for information processing, p. 344. Finansy i statistika, Moskow (2002). (in Russian)
20. Shin, Y., Ghosh, J.: Efficient higher-order neural networks for classification and function approximation. The University of Texas at Austin (1995). https://www.researchgate.net/publication/2793545_Efficient_Higher-order_Neural_Networks_for_Classification_and_Function_Approximation. Accessed 28 Jan 2017
21. Shevelev Ju, P.: Discrete mathematics. Part 1: The theory of sets. Boolean algebra (Automated learning technology "Symbol"), p. 118. TUSUR University, Tomsk (2003). (in Russian)
22. Omondi, A., Premkumar, B.: Residue number systems: theory and implementation, p. 312. Imperial College Press, London (2007)

Adaptive Gateway Element
Based on a Recurrent Neurodynamical Model

Yury S. Prostov$^{(\boxtimes)}$ and Yury V. Tiumentsev

Moscow Aviation Institute (National Research University), Moscow, Russia
prostov.yury@yandex.ru

Abstract. Dynamic model of a recurrent neuron with a sigmoidal activation function is considered. It is shown that with the presence of a modulation parameter its activation characteristic (dependence between input pattern and output signal) varies from a smooth sigmoid-like function to the form of a quasi-rectangular hysteresis loop. We demonstrate how a gateway element can be build using a structure with two recognizing neurons and one output neuron. It is shown how its functional properties change due to changes in the value of the modulation parameter. Such gateway element can take the output value based on a weighted sum of signals from the recognizing neurons. On the other hand it can perform a complex binary-like calculation with the input patterns. We demonstrate that in this case it can be used as a coincidence detector even for disjoint-in-time patterns. Futhermore, under certain extreme conditions it can be triggered even if only the one input pattern was recognized. Also the results of numerical simulations presented and some directions for further development suggested.

Keywords: Neurodynamic model · Hystersis · Adaptiveness

1 Introduction

There are a number of challenges in the field of development of an intelligent control systems for highly autonomous robotic systems such as unmanned aerial vehicles. One of these challenges is related to a mechanism that provides the control system with the ability to accumulate the experience from processed data and apply it in further. In other words, it is the task of developing a model with unsupervised or semi-supervised online learning algorithm which can work effectively in a dynamic and uncertain environment under a condition of limited computing resources.

Related and similar tasks are already successfully solved by methods from the field of machine learning such as incremental learning models [1,2]. They can be applied to an environment in which there are some types of uncertainty and noise or other difficulties. Almost all of them are based on feedforward and simple recurrent architectures which means that they can not reliably maintain activity

© Springer International Publishing AG 2018
B. Kryzhanovsky et al. (eds.), *Advances in Neural Computation, Machine Learning, and Cognitive Research*, Studies in Computational Intelligence 736,
DOI 10.1007/978-3-319-66604-4_5

of neurons without an external signal. But from our point of view this is necessary for context dependent recognition and learning [3]. The LSTM model [4] is the most suitable in this case because it has a special context cell to maintain internal neural network activity. But learning algorithms of this model are based on supervised techniques which is unacceptable in our problem.

We previously proposed the concept of a neural network model [3] as one of the possible approaches to solving the problem outlined above. Our model divided into two parts: working network that performes pattern recognition and learning as well as auxiliary network that evaluates the first one and produces the value of the modulation parameter. An important feature of neuron model used in a working network is that its activation characteristic (i.e. the dependence between an input pattern and an output value) can contain a hysteresis loop [5] under certain values of the modulation parameter as described in [6,7]. And as shown in [8,9] the presence of a hysteresis loop in an activation function is related to robust implementations of some working memory models. But at the same time activation characteristic will have the form of a smooth curve under other values of the modulation parameter. Thus, we can change the behaviour of neurons from gradual to trigger mode and can use this property to implement a gateway element with some interesting features.

2   Neuron Model

A neuron model used in this paper differs in some details from the one which was described in the related article [7]. Namely, in this article the activation function was replaced by a sigmoidal function and the threshold parameter was moved into a weights vector as one of the coefficients. As a result, model became as follows:

$$
\begin{cases}
du/dt & = \alpha y + i(\mathbf{w}, \mathbf{x}) - \mu u, \\
y & = f(h(u, \theta)),
\end{cases}
\tag{1}
$$

where $u \in \Re$ is a potential variable, $y \in [0; 1]$ is an output variable, $\mathbf{w} \in \Re^M$ is a weights vector, $\mathbf{x} \in [0; 1]^N$ is an input vector, $\alpha \in [0; +\infty)$ is a recurrent connection weight, $\mu \in (0; 1)$ is a potential dissipation parameter, $\theta \in (0; +\infty]$ is a modulation parameter, $i(\mathbf{w}, \mathbf{x})$ is an external excitation function (in the following we will omit the arguments for brevity) which can be specified as a scalar product or as a Gaussian radial basis function or as any other distance measure function, $h(u, \theta) = u/\theta$ is a potential modulation function, $f(z) = \sigma(z - \Delta)$ is a sigmoidal activation function with $\Delta = 3.0$. Also we assume that the values of parameters α and μ are fixed and selected in advance while the value of a modulation parameter θ is changeable during model operating.

It can be shown that the value of variable y will converge exponentially to some stable equilibrium point y^* of the dynamic system (1). To find these points we need to rewrite equations (1) as follows:

$$
F(y) = du/dt = \alpha y + i - \mu \theta g(y)
\tag{2}
$$

where $g(y) = \Delta + \log(y/(1-y))$ is the function inverse to the function f. In this case the equilibrium points can be determined from the condition $F(y^*) = 0$. Moreover, equilibrium point y_j^* will be stable if $F'(y_j^*) < 0$ and unstable if $F'(y_j^*) > 0$, otherwise, additional analysis will be required. It should be noted that there is no analytical solution and therefore this equation must be solved either graphically as shown on Fig. 1 or using numerical methods.

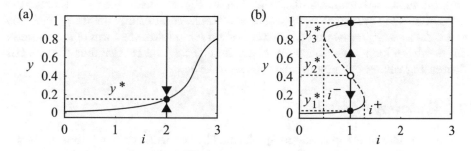

Fig. 1. Graphical solutions of a model (1): (a) the stable point y^* in the case of monostability; (b) the stable points $y_{1,3}^*$ and the unstable point y_2^* in the case of bistability

At the same time bifurcation analysis can be performed analytically based on the Eq. (2). It can be shown that a model (1) has cusp catastrophe [10] by parameters θ and i. Corresponding pitchfork bifurcation at the point $\theta = \alpha/4\mu$ shown in Fig. 2a where the values of parameter i at each point was choosen to get a symmetrical curve. In the case of $\theta \geq \alpha/4\mu$ there exist only one stable equilibrium point and activation characteristic function has the form of a sigmoidal curve as shown in Fig. 1a. Moreover, the slope of this curve decreases as the value of the parameter θ increases.

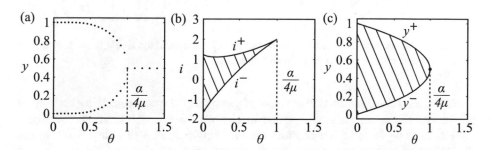

Fig. 2. The properties of the dynamic system (1): (a) a pitchfork bifurcation at the point $\theta = \alpha/4\mu$; (b) the dependence of the thresholds $i^{\pm}$ on the modulation parameter θ (shaded region corresponds to a bistability area); (c) the boundary between areas of a stable (not shaded region) and unstable (shaded region) points

In the case of $\theta < \alpha/4\mu$ a bistability region arises and it corresponds to the range $i \in (i^-; i^+)$ as shown in Fig. 1b. As we can see increasing of the parameter i value leads to abrupt change of the output value y at the point i^+ and a similar abrupt change occurs at the point i^- during its decreasing. It can be shown that these threshold values $i^\pm$ are determined by the extremes of the Eq. (2) and can be evaluated as follows: $i^\pm = -\alpha y^\pm + \mu\theta g(y^\pm)$ where $y^\pm = 0.5 \mp \sqrt{0.25 - \mu\theta/\alpha}$. Figure 2b shows the dependence between these thresholds and modulation parameter θ. As shown in Fig. 2c the values $y^\pm$ themselves determine the region $(y^-; y^+)$ where the stable equilibrium points y^* can not exist. As a result, activation characteristic function takes the form of a hysteresis curve with a loop which becomes closer and closer to a rectangular shape with decreasing value of the modulation parameter θ.

3   Gateway Model

Let us consider now a gateway model formed by connecting the neurons as shown in Fig. 3a. As a result, overall gateway state will be described as follows:

$$\begin{cases} du_k/dt & = \alpha y_k + i_k - \mu u_k, \\ y_k & = f(h(u_k, \theta)), \end{cases} \tag{3}$$

where the first and second ($k = 1, 2$) neurons process input patterns from two different data channels with appropriate external excitation values $i_{1,2}$ and the third one ($k = 3$) generates the gateway output signal $o = y_3$ using excitation value $i_3 = \beta y_1 + \beta y_2$.

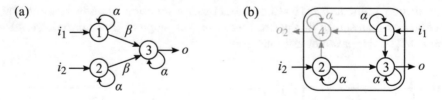

Fig. 3. Architecture: (a) gateway element; (b) computational element for future research based on the model of gateway element

Consider the case when the value of parameter θ corresponds to the region of monostability. As we noted early, in this case activation characteristic of a neuron has the form of a sigmoidal curve. Denote it as a function $\phi_\theta(i)$ where the subscript emphasizes the dependence between the slope of this curve and the value of modulation parameter θ. Then the value of gateway output signal can be represented as $o = \phi_\theta(i_3)$ where $i_3 = \beta \cdot (\phi_\theta(i_1) + \phi_\theta(i_2))$. Thus, the external signals i_1 and i_2 will be transfered by the gateway element in the form of nonlinear weighted sum and the amplitudes of these nonlinear transformations are controlled by the value of modulation parameter θ.

In the case of bistability the value of neuron output variable y takes values from the neighborhoods of points 0 (inactive state) and 1 (active state) as shown in Fig. 2c, i.e. $y \in O_\theta^+(0)$ and $y \in O_\theta^-(1)$ where the subscript emphasizes the dependence between the width of neighborhoods and the value of modulation parameter θ. In this case we can conclude that the value of $i_3 \in O_\theta^-(2\beta)$ if both of values i_1 and i_2 overcome the threshold value i^+ and $i_3 \in O_\theta(\beta)$ if only the one of them overcome the threshold value and otherwise $i_3 \in O_\theta^+(0)$. But the gateway output o can be in active state only when the value of i_3 overcome the threshold i^+. So, for a some fixed range of modulation parameter θ values we can choose the value of parameter β that satisfy to inequality $z_1 < i^+ < z_2$ for $\forall z_1 \in O_\theta(\beta)$ and $\forall z_2 \in O_\theta(2\beta)$. In this case the gateway output o will be active only when both input patterns from data channels are recognized. In other words, the gateway element will become a coincidence detector. But on the other side, we also can choose the value of parameter β which will admit an activation of the gateway element even if input pattern from only the one channel recognized.

Also note the extreme condition when the value of threshold parameter i^- falls below zero as shown in Fig. 2b. In this case the neurons that had previously passed into the active state can remain active even if there is no input signal. As a result, the gateway element can determine a coincidence by disjoint-in-time patterns due to self-sustained activity of recognizing neurons.

We performed numerical simulation to confirm the results obtained above with the following parameters: $\mu = 0.75$, $\alpha = 3.0$ and $\beta = 0.5$. During the simulation we explicitly changed the values of external excitation signals $i_{1,2}$ as well as the value of modulation parameter θ. As shown in Fig. 4 the results of simulation meet with our expectations. The case of performing a nonlinear weighted summation corresponds to the time interval $[t_1; t_2]$. The case of input patterns coincidence detection corresponds to the time interval $[t_3; t_4]$ and the special case of coincidence detection for disjoint-in-time patterns corresponds to the time interval $[t_5; t_6]$.

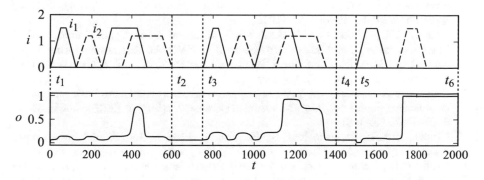

Fig. 4. Results of performed numerical simulation with different values of the external excitation signals $i_{1,2}$ and different values of the modulation parameter θ

4 Conclusions

We demonstrated that activation characteristic of the described neurodynamical model of neuron can vary from a smooth sigmoid-like function to the form of a quasi-rectangular hysteresis loop. It was shown how these changes are controlled by the value of modulation parameter θ and how this parameter is related with other parameters of the model.

Also we demonstrated how a gateway element can be build using the described neuron model. It was shown that for the certain range of modulation parameter values the gateway element transfers the input signals as a nonlinear weighted sum but for the other range of modulation parameter values it begins to perform a binary-like calculation as a complex coincidence detector with additional functional features.

As shown in Fig. 3b further research is related to the development of a computational element that would learn to associate the ascending and descending data streams in neural network based on the results obtained here for the model of gateway element.

References

1. Ditzler, G., Roveri, M., Alippi, C., Polikar, R.: Learning in nonstationary environments: a survey. IEEE Comput. Intell. Mag. **10**(4), 12 (2015). doi:10.1109/MCI. 2015.2471196
2. Gepperth, A., Hammer, B.: Incremental learning algorithms and applications. In: European Symposium on Artificial Neural Networks (ESANN) (2016)
3. Prostov, Y.S., Tiumentsev, Y.V.: Multimodal associative neural network with context-dependent adaptation. In: Abstracts of the 12th International Conference "Aviation and Cosmonautics - 2013", MAI (NRU), Moscow, pp. 619–620 (2013). (in Russian)
4. Hochreiter, S., Schmidhuber, J.: Long short-term memory. Neural Comput. **9**(8), 1735 (1997). doi:10.1162/neco.1997.9.8.1735
5. Krasnosel'skii, M.A., Pokrovskii, A.V.: Systems with Hysteresis. Springer Science & Business Media, Heidelberg (2012)
6. Prostov, Y.S., Tiumentsev, Y.V.: A study of neural network model composed of hysteresis microensembles. In: Proceedings of XVII All-Russian Scientific Engineering and Technical Conference "Neuroinformatics-2015", vol. 1, NRNU MEPhI, Moscow, pp. 116–126. (2015). (in Russian)
7. Prostov, Y.S., Tiumentsev, Y.V.: A hysteresis micro ensemble as a basic element of an adaptive neural net. Opt. Mem. Neural Netw. **24**(2), 116 (2015). doi:10.3103/ S1060992X15020113
8. Koulakov, A.A., Raghavachari, S., Kepecs, A., Lisman, J.E.: Model for a robust neural integrator. Nat. Neurosci. **5**(8), 775–782 (2002). doi:10.1038/nn893
9. Brody, C.D., Romo, R., Kepecs, A.: Basic mechanisms for graded persistent activity: discrete attractors, continuous attractors, and dynamic representations. Curr. Opin. Neurobiol. **13**(2), 204–211 (2003). doi:10.1016/S0959-4388(03)00050-3
10. Zeeman, E.C.: Catastrophe Theory: Selected Papers, 1972–1977. Addison-Wesley, Reading (1977)

Common Sense Knowledge in Large Scale Neural Conversational Models

D.S. Tarasov[✉] and E.D. Izotova

Meanotek AI Research, Sibirsky Trakt 34, Kazan, Russian Federation
dtarasov@meanotek.io

Abstract. It was recently shown, that neural language models, trained on large scale conversational corpus such as OpenSubtitles have recently demonstrated ability to simulate conversation and answer questions, that require common-sense knowledge, suggesting the possibility that such networks actually learn a way to represent and use common-sense knowledge, extracted from dialog corpus. If this is really true, the possibility exists of using large scale conversational models for use in information retrieval (IR) tasks, including question answering, document retrieval and other problems that require measuring of semantic similarity. In this work we analyze behavior of a number of neural network architectures, trained on Russian conversations corpus, containing 20 million dialog turns. We found that small to medium neural networks do not really learn any noticeable common-sense knowledge, operating pure on the level of syntactic features, while large very deep networks shows do posses some common sense knowledge.

Keywords: Deep highway networks · Conversational modeling · Information retrieval

1 Introduction

It was recently shown [1] that large-scale heterogenous dialog corpus can be used to train neural conversational model, that exhibits many interesting features, including capabilities to answer common-sense questions. For example, neural network model can tell that dog have four legs, and usual color of grass is green, even though these question/answer pairs do not explicitly exists in the dataset. This raises a question if such model can learn implicit ontology from conversations. If true, such models can be applied to the tasks outside of dialog modeling domain, such as information retrieval and question answering.

Unfortunately, this property has not received yet sufficient attention. Recent research on neural conversational models have been focused on incorporating longer context [2, 3], dealing with generic reply problem [4], incorporating attention and copying mechanism [5]. Attempts to connect neural conversational models to external knowledge bases were also made [6], however, we are not aware of any papers that investigated nature of knowledge that can be stored in neural network synaptic weights.

© Springer International Publishing AG 2018
B. Kryzhanovsky et al. (eds.), *Advances in Neural Computation, Machine Learning, and Cognitive Research*, Studies in Computational Intelligence 736,
DOI 10.1007/978-3-319-66604-4_6

In this work, we investigate the possibility of using large dialog corpus to train semantic similarity function. We train a number of neural network architectures, including recently proposed deep highway neural network model [7] on large number of dialog turns, extracted from both Russian part of OpenSubtitles database [8] and data collected from publicity available books in Russian, totaling 20 millions dialog turns. The training goal was to classify if sentence represent a valid response to previous utterance or not.

We found that smaller neural network models can learn general similarity function in sentence-space. This function performance is superior to simple neural bag of words models in selecting proper dialog responses and finding sentences, relevant to the query. However, these networks don't incorporate any meaningful knowledge about the world.

Large neural networks seem to incorporate some common-sense knowledge to semantic similarity function, as demonstrated by reranking possible answers to various common-sense and factoid questions.

2   Methods and Algorithms

2.1   Datasets

Russian part of OpenSubtitles database was downloaded from http://opus.lingfil.uu.se/. OpenSubtitles [8] is a large corpus of dialogs, consisting of movie subtitles. However, the data in this corpus is much smaller (about 10 M dialog turns after deduplication) then its English counterpart. OpenSubtitles is also very noise dataset, because it contains monologues, spoken by the same character, that are impossible to separate from dialogues, and also dialog boundaries are unclear.

To extend the available data for this work, we used Russian web-site lib.ru and mined publicly available fiction books for conversations of book characters. A heuristic parser was written to extract dialog turns from book texts. 10 M dialog turns was mined by this approach, resulting in total corpus size of 20 M dialog turns.

2.2   Neural Network Architectures

The structure of models, used for this work is shown on the Fig. 1. A number of specialized architectures were proposed for sentence matching task [9], including convolutional and LSTM models.

Overall, our model consists of two encoder layers that compute representations of source sentences, one or more processing layers stacked on top of each other and output layer, consisting of a single unit that outputs the probability of response being appropriate to context. In this work we tested two types of encoders LSTM-based encoder along with simpler fully connected encoder.

Neural bag of words (NboW) model is a fixed length representation xf obtained by summing up word vectors in the text and normalizing result (by multiplying by $1/|xf|$). This model was used as a baseline in [9].

2.3   Word Vectors

Real-valued embedding vectors for words were obtained by unsupervised training of Recurrent Neural Network Language Model (RNNLM) [10] over entire Russian Wikipedia. Text was preprocessed by replacing all numbers with #number token and all occurrences of rare words were replaced by corresponding word shapes.

3   Results and Discussion

3.1   Reply Selection Accuracies

Table 1 reports response selection accuracies for three different models on the test set, consisting of 10,000 contexts. For each context 4 random responses were given to classifier to rank along with "correct" (actual response from dataset).

Table 1. Model accuracies in selecting right context/response pairs

Model	Accuracy
Random baseline	19.7
Neural bag of word encoder with 1 fully connected processing layer	21.2
Fully connected encoder with 1 fully connected processing layer	37.8
Fully connected encoder with 4 highway processing layers	41.1
LSTM encoder with 4 highway processing layers	39.3

Two findings are particularly surprising. First, NboW model did not achieved any significant improvements over random baseline, in contrast with results reported in [9] for matching English twitter responses. This result might be due to the fact that our corpus is much larger (about 10 times) and much more noisy. Second, LSTM encoder actually performs worse than simple fully connected encoder, and it is also much slower. This is interesting, because fully-connected encoders with zero-padded sentences are not commonly evaluated for such tasks, because they are assumed to be bad models, because of their potential to overfit the data. However, with a special case of conversation, where most responses are small in size, and given a lot of data, apparently fully-connected encoders could be usable option.

Another interesting point here is that we observed that small model with 1 processing layer also scored 29.8 on the task of matching English sentences using pre-trained word vectors for English language, without training the network itself on English data. This result indicates that small models actually learn some language-independent generic similarity function that operate on word vectors and not involve deeper understanding of the content.

3.2   Factoid Answer Selection from Alternatives

To evaluate model capability for question answering, we designed a test set of 300 question-answers pairs, using search engine snippets as candidate answers. The task was to select snippet, containing the correct answer (all snippets were first evaluated by human, to asses if they contain necessary answers). Top 10 snippets were selected for evaluation for each question. Table 2 summarizes results of all models.

Table 2. Accuracies on factoid question answering

Model	Percent of correct answers
First snippet baseline	30.3%
Fully connected encoder with 1 fully connected processing layer	36.2%
Fully connected encoder with 4 highway processing layers	34.7%
LSTM encoder with 4 highway processing layers	32.0%

For this task, a model with one processing layer demonstrated best results. Overall, improvements over were small, probably because search engine snippets already represent strong baseline. Manual inspection of ranking results revealed, that improvements were due to models capacity to distinguish between snippets that contained answers and snippets that were just copies of the questions (see Table 3).

Table 3. Example ranking of candidate snippets for the question "сколько звезд на небе" (How many stars are there in the sky?)

Answer text	Answer ranking
В ясную погоду на небе видно около 3000 звезд ("*on clear whether, about 3000 stars can be seen at the sky*")	0.76
Сколько же звезд на небе? ("*How many stars are in the sky?*")	0.68
На этой странице вы узнаете, сколько звезд на самом деле видно на небе ("*On this page you will learn how many stars can be seen at the sky*")	0.55

We therefore conclude, that model can use sentence structure to decide if it can be viewed as appropriate answer or not.

3.3 Common Sense Questions

Finally, to test models capacity to understand the world, we prepared a set of 100 common-sense questions, like "what is the color of the sky?", "what pizza is?". Like in previous setup, we evaluate model capability to choose correct answer out of 5 options. Results are summarized in Table 4.

Table 4. Accuracies on multiple-choice common-sense questions

Model	Result
Random baseline	19.5%
Fully connected encoder with 1 fully connected processing layer	20.3%
Fully connected encoder with 4 highway processing layers	26.5%
LSTM encoder with 4 highway processing layers	19.8%

Only deep model with fully-connected encoder demonstrated some understanding of common sense questions above random baseline and even here results are generally poor. Table 5 shows example rankings of answers to a typical question by best model.

Table 5. Example ranking of candidate answers for common sense questions

Что такое собака? *What dog is?*	Что такое автомобиль? *What automobile is?*	где живет человек? *Where does human live?*	Какого цвета земля? *What is the color of the ground?*
0.71 животное (animal)	0.84 мотор (motor)	0.844 нора (burrow)	0.87 зеленая (green)
0.49 растение (plant)	0.70 механизм (mechanism)	0.841 дом (house)	0.86 желтая (yellow)
0.48 концепция (concept)	0.67 животное (animal)	0.52 лес (forest)	0.83 бурая (brown)
0.45 планета (planet)	0.65 фонарь (lamp)	0.48 лужа (water pool)	0.82 черная (black)
0.43 механизм (mechanism)	0.59 квадрат (square)	0.42 джон (John)	0.79 белая (white)
0.38 фонарь (lamp)	0.54 дом (house)	0.41 мотор (motor)	0.70 дом (house)
0.36 дом (house)	0.46 планета (planet)	0.40 фонарь (lamp)	0,62 красная (red)
0.32 квадрат (square)	0.46 концепция (concept)	0.35 квадрат (square)	0.46 фонарь (lamp)
0.30 африка (Africa)	0.45 африка (Africa)	0.17 море (sea)	0.45 синяя (blue)
0.19 джон (John)	0.39 растение (plant)		0.16 джон (John)

Manual examination rankings revealed, that questions that concern relationships of two and more entities are more difficult to answer, compared to the questions related to the single entity (Table 5)

4   Conclusions

We found that large neural dialog models can learn some common-sense knowledge, although to the limited extent. There is, however, a room for improvement, because we found that even our large model did not significantly overfit the training set, and there is also a possibility for collecting more training data.

Another interesting finding is that our models learned to understand sentence structure of question/answer pairs and can select answers those structure is more likely to contain answers to the question.

Finally, we observed that simple encoder, based on fully-connected layer with padded input outperforms LSTM-based encoders both in computing speed and response selection accuracy. Further analysis is needed to understand the significance of this finding.

Subsequent work should probably include analysis of even larger models, and detailed analysis of what happens in encoding layers, to better understand how these models really operate and what they can do. Also, testing sets need to be expanded in both size and extend of coverage of various common-sense topics.

References

1. Vinyals, O., Le, Q.: A neural conversational model. arXiv preprint, arXiv:1506.05869 (2015)
2. Yao, K., Zweig, G., Peng, B.: Attention with intention for a neural network conversation model. arXiv preprint, arXiv:1510.08565 (2015)
3. Chen, X., et al.: Topic aware neural response generation. arXiv preprint, arXiv:1606.08340 (2016)
4. Li, J., Galley, M., Brockett, C., Gao, J., Dolan, B.: A diversity-promoting objective function for neural conversation models. arXiv preprint, arXiv:1510.03055 (2015)
5. Mihail, E., Manning, D.: A copy-augmented sequence-to-sequence architecture gives good performance on task-oriented dialogue. arXiv preprint, arXiv:1701.04024 (2017)
6. Ahn, S., et al.: A neural knowledge language model. arXiv preprint, arXiv:1608.00318 (2016)
7. Srivastava, R.K., Greff, K., Schmidhuber, J.: Highway networks. arXiv preprint, arXiv:1505.00387 (2015)
8. Tiedemann, J.: News from OPUS—a collection of multi-lingual parallel corpora with tools and interfaces. In: Nicolov, N., Bontcheva, K., Angelova, G., Mitkov, R. (eds.) Recent Advances in Natural Language Processing, pp. 237–248. John Benjamins Publishing Company, Amsterdam (2009)
9. Hu, B., Lu, Z., Li, H., Chen, Q.: Convolutional neural network architectures for matching natural language sentences. In: Advances in Neural Information Processing Systems, pp. 2042–2050 (2014)
10. Mikolov T., Karafiat M., Burget L., Cernocky J., Khudanpur S.: Recurrent neural network based language model. In: INTERSPEECH, pp. 1045–1048 (2010)

Applications of Neural Networks

Prospects for the Development of Neuromorphic Systems

Aleksandr Bakhshiev[1](✉) and Lev Stankevich[2]

[1] Russian State Scientific Center for Robotic and Technical Cybernetics,
Saint-Petersburg 195257, Russian Federation
alexab@rtc.ru
[2] Peter the Great Saint-Petersburg Polytechnic University,
Saint-Petersburg 195251, Russian Federation

Abstract. The article is devoted to the analysis of neural networks from the positions of the neuromorphic approach. The analysis allows to conclude that modern artificial neural networks can effectively solve particular problems, for which it is permissible to fix the topology of the network or its small changes. In the nervous system, as a prototype, the functional element - the neuron - is a fundamentally complex object, which allows implementing a change in topology through the structural adaptation of the dendritic tree of a single neuron. Promising direction of development of neuromorphic systems based on deep spike neural networks in which structural adaptation can be realized is determined.

Keywords: Neuromorphic system · Spike model · Neuron · Structural adaptation · Deep learning · Neuromorphic computing

1 Introduction

Currently, there are many poorly formalized problems that are badly solved by existing methods (detection and recognition of objects in conditions of significant data shortage, control of unstable systems, control of the behavior of mobile agents in a volatile environment, etc.).

One of the most promising common approaches to solving such problems is artificial neural networks (ANN), in particular, deep neural networks (DLN), which are now actively developing. This is due, in particular, with the advent of new hardware (NVIDIA graphics accelerators [1], specialized processors (BrainScaleS [2, 3], SpiNNaker [4], NIDA [5], DANNA [6], Neurogrid [7], IBM TrueNorth [8]), which allow efficient numerical calculations on the basis of the mathematical apparatus of the DLN, and the direction of neuromorphic systems, whose architecture and design are based on the principles of the work of the biological neural structures of the nervous system. This is a fairly broad interpretation, in which the deep learning fit well. Possible successes of neuromorphic systems are associated, first of all, with the biological plausibility of their basic neuron component and its hardware implementation. In this sense, some specialized processors (in particular, IBM TrueNorth) refer specifically to processors of the neuromorphic type.

© Springer International Publishing AG 2018
B. Kryzhanovsky et al. (eds.), *Advances in Neural Computation, Machine Learning, and Cognitive Research*, Studies in Computational Intelligence 736,
DOI 10.1007/978-3-319-66604-4_7

2   Overview of Deep Neural Network Architectures

Today, the practical application of neural networks is most intensively developed in the trend of deep learning.

There is a large number of networks within this trend [9]. The basic architectures, from which all the main implementations are obtained:

- Feed forward (FF) (Perceptron, Autoencoders [10]);
- Fully connected networks (FCN) (Markov Chain [11], Hopefield network [12], Boltzmann Machine [13];
- Convolutional neural networks (CNN) (LeNet [14], VGG-19 [15], Google Inception [16]);
- Recurrent neural networks (RCN) (LSTM [17], Deep Residual Network (ResNet) [18–20]);

There are separately presented architectures such as growing neural networks, in which the following widespread types can be distinguished:

- Networks based on Kohonen maps (SOM [21], ESOM [22], GHSOM [23], SOS [24]);
- SOINN, ESOINN [25];
- Neural Gas Network [26] and its derivatives GNG [27], IGNG [28], GCS [29], TreeGCS [30], PGCS [31] and others.

Relatively new works are devoted to the implementation of spiking neural networks, based on the above architectures [32–34]. The advantages of deep spiking neural networks are firstly declared in the significant energy savings in the case of hardware implementation.

If we consider the achievements of neural networks from the point of view of solving particular problems, great progress has been made in this direction. So, according to the results of the competition in recent years, DLN have been won in most computer vision tasks (pattern recognition, object detection, segmentation, etc.). It is important to note that such networks are effective in problems in which there are high local correlations in the input data.

Also, there is the big problem of combining a set of private solutions, formed by neural networks to solve common problems of controlling agent behavior in a complex environment. In other words, the solution, for example, of object detection problem, converts the space of high-dimensional input data into a space of low dimensionality of the classes of objects to be detected. If it is necessary to create a flexible control system for the behavior of the agent (robot) in a volatile environment, we are forced to operate with a number of such particular solutions. This naturally limits the agent in adaptability to changes in the environment. Part of this problem is solved in growing networks.

Despite the fact that ANN were originally based on the analogy with the nervous system, the majority of neural networks in their topology, training rules and principles of functioning as a whole is very different, and the trend away from biological likelihood is growing. In particular, the development of networks follows the path of increasing the number of layers, but not the complexity of the functional element of neural networks

- the neuron; and growing neural networks are based on the addition of neurons and layers, in contrast to change in the structure of a neuron dendritic tree in a biological system, where each dendrite provides complex information processing.

If we compare the known features of the nervous system and ANN (assuming that the advantages of the still disjointed architectures of ANN will be unified), then following table can be made (Table 1).

Table 1. Comparison of the features of artificial neural networks and the nervous system

System property	Artificial neural network	Nervous system
The complexity of the functional element	Low	High
The possibilities of structural adaptation of the network	The network topology is rigidly defined within the architecture. Topology can be changed block-wise using global optimization algorithms	Topology is partially defined by DNA, but low-level parts can change their function (solved tasks), at the initial stage of growth, and high-level parts always
The principle of remembering information in the network structure	Generalization of input data and reduction of the dimension of the problem. Formation of one (or a limited number) of output vectors	Generalization of input data and reduction of the dimension of the problem. Formation of a set of vectors of output data (work simultaneously in a set of contexts)
Method of network restructuration	Change the number of neurons in the layer, the number of layers, the number of neurons in the ensemble	Change in the structure of the neuron membrane (number and length of dendrites—generalizing elements, the number of synapses, the size of the neuron). Change in the number of neurons in the "layer"/ ensemble, the number of "layers"
Methods for parametrizing the network	Change in the weight of the neuron input	Change the size of the synapse

It seems promising to consider the possibility of complicating the model of the neural networks functional element with an emphasis on the possibilities of network structural adaptation, in the trend of the neuromorphic approach.

3   Neuron Models

There are many widespread models of neurons. By the level of abstraction, models can be divided into:

- Biological (biophysical)-models based on the modeling of biochemical and physiological processes, which, as a consequence, lead to a certain behavior of the neuron in certain modes of operation (the Hodgkin–Huxley model [35]).
- Phenomenological-models describing certain phenomena of the behavior of a neuron in certain modes of operation as a "black box" (the Izhikevich model [36]).
- Formal-models with the highest level of abstraction, describing only the basic properties of the neuron (formal neuron [37]).

Each model can correspond to several features from this classification. In the framework of ANN in general, and DLN, in particular, modifications of formal neuron models, with different activation functions (Sigmoid, hyperbolic tangent, ReLU and its derivatives [38]) are used. Spiking variations of deep networks basically contain such models of neurons as variations of the threshold integrator model [39], the Izhikevich model mentioned above.

One of the promising options for implementing the model of an element of neuromorphic systems is the phenomenological model of a dynamic spike neuron with the ability to describe the spatial structure of the dendritic apparatus [40]. This model allows us to describe the variable topology of a neural network, based on the principles of neural structure formation known from neurophysiology [41].

4   Discussion

The main feature of the nervous system, which is still not considered in the ANN archives, is a great potential in structural (topological) restructuring. Structural adaptation in the nervous system is largely based on the high complexity of a single element of the network - the neuron.

The analysis allows to identify the following areas of development of ANN in the framework of the neuromorphic approach:

- Complicating the neuron model, adding the possibility of describing the structure of the membrane (as generalizing and binding elements) of the neuron.
- Development of learning algorithms, taking into account the modification of the structure of the generalizing and binding elements of the neuron.
- Development of ANN architectures that allow training and data output simultaneously in multiple contexts.

References

1. Chetlur, S., Woolley, C., Vandermersch, P., Cohen, J., Tran J.: cuDNN: Efficient Primitives for Deep Learning arXiv:1410.0759v3 [cs.NE], 18 December 2014
2. Pfeil, T., Grübl, A., Jeltsch, S., Müller, E., Müller, P., Petrovici, M.A., Schmuker, M., Brüderle, D., Schemmel, J., Meier, K.: Six networks on a universal neuromorphic computing substrate. Front. Neurosci. 7(11) (2013). doi:10.3389/fnins.2013.00011
3. Schemmel, J., Bruderle, D., Grubl, A., Hock, M., Meier, K., Millner, S.: A wafer-scale neuromorphic hardware system for large-scale neural modeling. In: Proceedings of 2010 IEEE International Symposium on Circuits and Systems (ISCAS), pp. 1947–1950. IEEE (2010)

4. Furber, S.B., Lester, D.R., Plana, L., Garside, J.D., Painkras, E., Temple, S., Brown, A.D., et al.: Overview of the spinnaker system architecture. IEEE Trans. Comput. **62**(12), 2454–2467 (2013)
5. Schuman, C.D., Birdwell, J.D.: Variable structure dynamic artificial neural networks. Biol. Inspired Cognit. Archit. **6**, 126–130 (2013)
6. Schuman, C.D., Disney, A., Reynolds, J.: Dynamic adaptive neural network arrays: a neuromorphic architecture. In: Workshop on Machine Learning in HPC Environments, Supercomputing (2015)
7. Benjamin, B.V., Gao, P., McQuinn, E., Choudhary, S., Chandrasekaran, A.R., Bussat, J.-M., Alvarez-Icaza, R., Arthur, J.V., Merolla, P., Boahen, K., et al.: Neurogrid: a mixed-analog-digital multichip system for large-scale neural simulations. Proc. IEEE. **102**(5), 699–716 (2014)
8. Merolla, P.A., Arthur, J.V., Alvarez-Icaza, R., Cassidy, A.S., Sawada, J., Akopyan, F., Jackson, B.L., Imam, N., Guo, C., Nakamura, Y., et al.: A million spiking-neuron integrated circuit with a scalable communication network and interface. Science **345**(6197), 668–673 (2014)
9. Van Veen, F.: The Neural Network Zoo (2016). http://www.asimovinstitute.org/neural-network-zoo/. Accessed 16 Apr 2017
10. Kingma, D.P., Welling, M.: Auto-encoding Variational Bayes. arXiv preprint arXiv: 1312.6114 (2013)
11. Hayes, B.: First links in the Markov chain. Am. Sci. **101**(2), 252 (2013)
12. Hopfield, J.J.: Neural networks and physical systems with emergent collective computational abilities. Proc. Natl. Acad. Sci. **79**(8), 2554–2558 (1982)
13. Hinton, Geoffrey E., Sejnowski, Terrence J.: Learning and relearning in Boltzmann machines. Parallel Distrib. Process. **1**, 282–317 (1986)
14. LeCun, Y., et al.: Gradient-based learning applied to document recognition. Proc. IEEE. **86**(11), 2278–2324 (1998)
15. Simonyan, K., Zisserman, A.: Very deep convolutional networks for large-scale image recognition. In: Published as a Conference Paper at ICLR 2015
16. Szegedy, C., et al.: Rethinking the Inception Architecture for Computer Vision. arXiv: 1512.00567v3 [cs.CV], 11 December 2015
17. Hochreiter, S., Schmidhuber, J.: Long short-term memory. Neural Comput. **9**(8), 1735–1780 (1997)
18. He, K., et al.: Deep residual learning for image recognition. arXiv preprint arXiv:1512.03385 (2015)
19. Moniz, J., Pal, C.: Convolutional Residual Memory Networks. arXiv:1606.05262 [cs.CV], 14 July 2016
20. Targ, S., Almeida, D., Lyman, K.: Generalizing Residual Architectures. arXiv:1603.08029v1 [cs.LG], 25 March 2016
21. Kohonen, Teuvo: Self-organized formation of topologically correct feature maps. Biol. Cybern. **43**(1), 59–69 (1982)
22. Deng, D., Kasabov, N.: ESOM: an algorithm to evolve self-organizing maps from on-line data streams. In: Proceedings of the International Joint Conference on Neural Networks (IJCNN 2000), Como, Italy, 24–27 July 2000, vol. vi, pp. 3–8. IEEE computer society (2000)
23. Growing Hierarchical Self-Organizing Map (GHSOM), Dittenbach, M., Merkl, D., Rauber, A.: The growing hierarchical self-organizing map. In: Proceedings of the International Joint Conference On Neural Networks (IJCNN 2000), vol. VI, pp. 15–19

24. Self-Organizing Surfaces (SOS), Zell, A., Bayer, H., Bauknecht, H.: Similarity analysis of molecules with self-organizing surfaces—an extension of the self-organizing map. In: Proceedings of International Conference on Neural Networks, ICNN 1994, Piscataway, pp. 719–724 (1994)
25. Furao, S., Hasegawa, O.: An incremental network for on-line unsupervised classification and topology learning. Neural Netw. **19**(1), 90–106 (2006)
26. Martinetz, T., Schulten, K.: A "neural gas" Network Learns Topologies. Artificial Neural Networks, pp. 397–402. Elsevier, Amsterdam (1991)
27. Fritzke, B.: A growing neural gas network learns topologies. Adv. Neural. Inf. Process. Syst. **7**, 625–632 (1995)
28. Prudent, Y., Ennaji, A.: An incremental growing neural gas learns topologies. In: Neural Networks, IJCNN 2005 (2005)
29. Fritzke, B.: Growing cell structures- a self-organizing network for unsupervised and supervised learning. Neural Netw. **7**(9), 1441–1460 (1994)
30. Hodge, V., Austin, J.: Hierarchical growing cell structures: TreeGCS. In: IEEE TKDE Special Issue on Connectionist Models for Learning in Structured Domains
31. Vlassis, N., Dimopoulos, A., Papakonstantinou, G.: The probabilistic growing cell structures algorithm. Lecture Notes in Computer Science, vol. 1327, p. 649 (1997)
32. Hunsberger, E., Eliasmith, C.: Spiking deep networks with LIF neurons. arXiv:1510.08829v1 [cs.LG], 29 October 2015
33. Gavrilov, A., Panchenko, K.: Methods of learning for spiking neural networks. A survey. In: 13th International Scientific-Technical Conference APEIE–39281, At Novosibirsk, vol. 1, part 2, pp. 60–65 (2016)
34. Cao, Y., Chen, Y., Khosla, D.: Spiking deep convolutional neural networks for energy-efficient object recognition. Int. J. Comput. Vis. **113**(1), 54–66 (2015)
35. Hodgkin, A.L., Huxley, A.F.: A quantative description of membrane current and its application conduction and excitation in nerve. J. Physiol. **117**, 500–544 (1952)
36. Izhikevich E.M.: Simple model of spiking neurons. In: IEEE Transactions on Neural Networks. A Publication of the IEEE Neural Networks Council, vol. 14 (2003)
37. McCulloch, W.S., Pitts, W.: A logical calculus of the ideas immanent in nervous activity. Bull. Math. Biophys. **5**, 115–133 (1943)
38. Nair, V., Hinton, G.: Rectified linear units improve restricted Boltzmann machines. In: Proceedings of the 27th International Conference on Machine Learning (ICML 2010), Haifa, Israel, 21–24 June 2010
39. Burkitt, A.N.: A review of the integrate-and-fire neuron model: I. Homogeneous synaptic input. Biol. Cybern. **95**(1), 1–19 (2006)
40. Bakhshiev, A.V., Gundelakh, F.V.: Application the spiking neuron model with structural adaptation to describe neuromorphic systems. Procedia Comput. Sci. **103**, 190–197 (2017)
41. Nicholls, J.G., Martin, A.R., Fuchs, P.A., Brown, D.A., Diamond, M.E., Weisblat, D.A.: From Neuron to Brain. Sinauer Associates Incorporated, Sunderland (1999)

Pulse Neuron Learning Rules for Processing of Dynamical Variables Encoded by Pulse Trains

Vladimir Bondarev$^{(\boxtimes)}$

Sevastopol State University, 299053 Sevastopol, Russian Federation
bondarev@sevsu.ru

Abstract. The paper deals with a model of pulse neural network that is applicable for solving of various tasks of processing sensory information. These tasks relate to dynamical variables processing. The distinctive feature of the problem statement is that dynamical variables are represented by pulse (spike) trains. We propose two supervised temporal learning rules for pulse neural network executing the required linear dynamic transformations of variables represented by pulse trains. To generate the required output of the network model we used a reference system with desired properties. The rules minimize the difference between the actual and required pulse train in a local window. The first temporal learning rule was named WB-FILT as it uses the filtered values of errors between binary vectors representing the desired and actual pulse sequences. The second rule was named WB-INST as it uses instantaneous value of the error, which is the difference of the desired and the actual elements of binary vectors. We demonstrated rule's properties by computer simulation of the mappings of the regular and the dynamical pulse trains. It has been shown that proposed rules are able to configure the simple network that implements a linear dynamic system.

Keywords: Pulse neuron · Pulse train · Supervised learning · Dynamic system

1 Introduction

Now much attention is paid to the pulse neural networks (PNN) for processing of dynamical variables [1, 2]. In PNN the dynamical variables are encoded by pulse (spike) trains. Development of supervised learning rules for functional PNN which implements the required processing of dynamical variables during the mapping process of the input pulse trains to the desired output pulse trains is considered as an important problem in neuroinformatics [3].

Various temporal supervised learning rules providing the desirable mappings of the pulse trains and using precise time of pulses are proposed in [4–7]. However, in most cases they are oriented on pattern classification problems and are not aimed to the direct application in adaptive real-time systems where processing of the dynamical variables represented by the multi-pulse trains is required.

The vector-matrix digital model of the pulse neuron (PN) and the supervised learning rule for real-time adaptive signal processing were proposed in [8, 9].

© Springer International Publishing AG 2018
B. Kryzhanovsky et al. (eds.), *Advances in Neural Computation, Machine Learning, and Cognitive Research*, Studies in Computational Intelligence 736,
DOI 10.1007/978-3-319-66604-4_8

The purpose of this paper is the extension of the scope of the PN vector–matrix model [8, 9] that provides direct realization of the required linear transformations of dynamical variables based on the input and output pulse sequences of the PN.

2 Problem Formulation

We will consider the adaptive modeling scheme of the linear dynamic system appearing as a reference system which performs the required linear transformation (mapping) of the input dynamical variable $u(t)$ to the output variable $y_d(t)$ represented (encoded) by means of desired pulse sequence $s_d(t)$. We want to construct the PNN model which reproduces the dynamics of the reference system based on the desired (required) pulse train $s_d(t)$.

To solve the problem, we will use the multi-input PN model that was considered in [8]. It is assumed that bipolar input pulse trains $u_i(t)$ generated by the encoding presynaptic neurons arrive at inputs of the PN linear filters with pulse responses $h_i(t)$. Filter reactions $x_i(t)$ are weighted with synaptic weights w_i and summarised to form the summary postsynaptic potential $y_o(t)$ of the PN. If the integral of the module of $y_o(t)$ exceeds a threshold then an output pulse of the PN with the sign corresponding to the sign of $y_o(t)$ is emitted and integrator state is nullified. The specified chain of the conversions corresponds to the LIF-neuron.

If we calculate the values of $y_o(t)$ at discrete time $t_n = n\Delta t$, where Δt is a time sampling step, then [8, 9]

$$y_o(n) = \mathbf{w}^T \mathbf{x}(n), \quad x_i(n) = \mathbf{b}_i^T(n)\mathbf{h}_i, \tag{1}$$

where $\mathbf{w}^T = (w_0, w_1, \ldots, w_{I-1})$ is synaptic weight vector, $\mathbf{b}_i^T(n)$ is sliding binary vector whose elements are equal to signs of the input pulses at time moments t_n, $\mathbf{h}_i$ denotes the impulse response vector $\mathbf{h}_i = (h_i(0), h_i(1), \ldots, h_i(K-1))^T$. In this case, we can use the supervised learning rule in the form of Widrow-Hoff [9]:

$$\Delta \mathbf{w}(n) = \mu \mathbf{x}(n)e(n), \tag{2}$$

Where μ is a learning rate, $e(n) = y_d(n) - y_0(n)$ is an error.

The rule (2) assumes that PN input signals are pulses and the output signal of the PN is represented by sample values of the dynamical variable $y_o(t)$. Therefore, the rule (2) cannot be used directly for training of a PNN where not only input signals, but also output signals are represented by pulse sequences.

We will derive the supervised PN learning rules for a case when the required output signal $y_d(t)$ of the reference dynamic system and the actual output signal $y_o(t)$ of the PN model (1) are represented by the pulse trains. We will call such rules that are driven directly by the time of pulses as temporal rules.

3 Temporal Learning Rules of the Pulse Neuron

In order to calculate the error $e(n)$ we will use the known similarity measures of the pulse trains [10, 11]. The most often used measure convolves of the pulse trains with some positive smooth localized kernel $h_r(t)$. In accordance with (1) the convolution of an actual output pulse train $s_o(t)$ and desired pulse trains $s_d(t)$ with a kernel $h_r(t)$ can be written as follows

$$\tilde{y}_o(n) = \mathbf{b}_o^T(n)\mathbf{h}_r, \; \tilde{y}_d(n) = \mathbf{b}_d^T(n)\mathbf{h}_r, \tag{3}$$

where $\mathbf{h}_r$ is the vector of samples of a kernel $h_r(t)$, $\mathbf{b}_o^T(n)$ and $\mathbf{b}_d^T(n)$ are the binary sliding vectors containing of M elements and corresponding to the pulse sequences $s_o(t)$ and $s_d(t)$. Length M is selected considering dynamics of the reference system and the processed signals. In fact, binary vectors fix some temporal prehistory of pulses.

Variables $\tilde{y}_o(n)$ and $\tilde{y}_d(n)$ can be interpreted as the result of conversions of dynamical variables $y_o(n)$ and $y_d(n)$ to the pulse trains s_o and s_d, and then back to the origin form for the purpose of restoration of these variables from the pulse sequences. If we perform replacement of $y_o(n)$ and $y_d(n)$ by variables $\tilde{y}_o(n)$ and $\tilde{y}_d(n)$ in the learning rule (2) we will derive the temporal learning rule

$$\Delta \mathbf{w}(n) = \mu \mathbf{x}(n)\left[(\mathbf{b}_d^T(n) - \mathbf{b}_o^T(n))\mathbf{h}_r\right]. \tag{4}$$

Having compared (2) and (4), we conclude that the error $e(n)$ in the rule (4) corresponds to the difference of binary vectors representing the desired and actual pulse sequences. At the same time this error is smoothed by a window (by a filter) with weights $\mathbf{h}_r$. We will name this temporal rule WB-FILT, as it compares filtered binary vectors (by analogy with [7]).

The window $\mathbf{h}_r$ is often selected so that the pulses (elements of binary vectors) which were formed later will have the greater weight. If the length of the window is restricted to a single sample then from (4) we derive the simple learning rule

$$\Delta \mathbf{w}(n) = \mu \mathbf{x}(n)(b_d(n) - b_0(n)), \tag{5}$$

where $b_d(n)$ and $b_o(n)$ are the elements of binary vectors. This temporal rule uses the instantaneous value of the error, which is equal to the difference of binary vectors elements. Therefore, we will name it WB-INST (by analogy with [7]).

4 Computer Simulation

During the simulation, the simple model of bipolar IF-neuron with single input was used as the model of an encoding neuron. The encoding neuron converts an input signal $u(t)$ to a pulse train. This pulse train simultaneously arrives to all inputs of the PN. Pulse responses of the PN filters were identical in the form, but shifted in time for the sampling step, i.e. $h_i(t) = \exp(-(t - i\Delta t)/\tau_s)$, where τ_s is the time constant.

To keep the shape of signals the finite symmetric exponent was used as a kernel function. The kernel function was shifted for the half of its length to provide the linear phase characteristic. Such kernel function creates the time delay equal to $(M-1)/2$ (if M is odd) that requires the correction of the rule (4):

$$\Delta\mathbf{w}(n) = \mu\mathbf{x}(n - (M-1)/2)\big[\big(\mathbf{b}_d^{\mathrm{T}}(n) - \mathbf{b}_o^{\mathrm{T}}(n)\mathbf{h}_r\big)\big]. \tag{6}$$

In the first computational experiment ($I = 401, K = 64, M = 129, \Delta t = 0.5$ ms), we run the training process to map the regular input pulse train with the period of 12.5 ms to the desired pulse train with the period of 20.5 ms. The specified pulse sequences were created by the encoding neurons when their inputs are constant signals with amplitude $u(t) = 0.08$ and $y_d(t) = 0.05$. It provided one pulse within the significant duration of the pulse response $h_i(t)$. In this case, the filter reactions to pulses in the separate channels of PN are not accumulated. It allows tracing the learning dynamics of PN visually.

The actual output pulse train in the form of a raster and the diagram of the mean-square error after training of the PN with the help of WB-INST rule (5) are illustrated in Fig. 1. The raster was created from the actual pulse train by cutting it into segments.

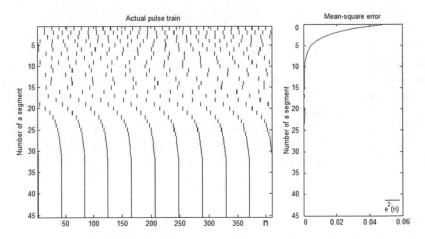

Fig. 1. Results of the mapping of the regular pulse sequences

Similar results also turn out using the WB-FILT rule (6). However, in case of a mapping of the regular pulse trains the WB-FILT rule provides faster convergence in comparison with the WB-INST rule due to averaging of the error $e(n)$ by $\mathbf{h}_r$.

In the second experiment ($I = 2001, K = 10, M = 65, \Delta t = 10$ ms), the training of a mapping of the dynamical pulse sequences was carried out. We want to build a PNN with the dynamics defined by the dynamics of the reference system which implements double integration of the dynamical input variable $u(t)$. The similar problem arises in the case of signal processing of accelerometers [12].

During the training, the input signal $u(t)$ equal to the sum of sine signals with the multiple frequencies was applied to the input of the encoding neuron, and the corresponding desired (reference) signaly $_d(t)$ arrived at the input of other encoding neuron. The desired output signal $y_d(t)$ is calculated with the help of normalized values of frequency response of the reference double integrator [12].

The distributions of the weight vector elements after training of PN are shown in Fig. 2. Interpreting **w** as a pulse response, it is possible to obtain the frequency response of the PNN model which corresponds to the double integrator in the bandpass range (curves 2 and 3). The frequency response of the reference double integrator (curve 1) was set in 30 uniformly distributed frequency points. Figure 2 shows that the mean square of the error $e(n)$ is decreasing with growth of n and the frequency response of the synthesized PNN model approximates the frequency response of the reference double integrator well. Pulse periodic behavior of the error is explained by periodicity of used signals.

It is interesting to note that despite the differences in the nature of elements distribution of the vector **w** for two rules (5) and (6) the frequency responses obtained with their help are the very close (curves 2 and 3).

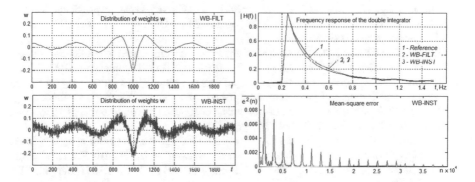

Fig. 2. Results of the mapping of the dynamical pulse sequences

5   Conclusions

The presented temporal supervised learning rules WB-INST and WB-FILT are applicable for using in digital adaptive systems with the reference PNN that performs the required linear transformations of the dynamical variables represented by pulse sequences.

The quantitative changes of synaptic weights are proportional to an error and reactions of the PN synaptic connections to input pulses. In such common formulation, the proposed temporal learning rules are similar to the known rules: ReSuMe [4], SPAN [5], PSD [6], INST and FILT [7].

However, an important distinction of the proposed temporal learning rules is that they are formulated in the discrete time in a general view. It allows deriving further variations of these rules oriented on specifics of processing tasks of dynamical variables. In addition, the offered PN model and rules due to the sparsity of binary vectors are quite effective from computational point of view.

References

1. Boerlin, M., Machens, C.K., Denève, S.: Predictive coding of dynamical variables in balanced spiking networks. PLoS Comput. Biol. **9**(11), 1–16 (2013). doi:10.1371/journal. pcbi.1003258
2. Memmesheimer, R.M., Rubin, R., Ölveczky, B.P., Sompolinsky, H.: Learning precisely timed spikes. Neuron **82**(4), 925–938 (2014). doi:10.1016/j.neuron.2014.03.026
3. Abbott, L.F., DePasquale, B., Memmesheimer, R.-M.: Building functional networks of spiking model neurons. Nat. Neurosci. **19**(3), 350–355 (2016)
4. Ponulak, F., Kasinski, A.: Supervised learning in spiking neural networks with ReSuMe: sequence learning, classification, and spike shifting. Neural Comput. **22**(2), 467–510 (2010). doi:10.1162/neco.2009.11-08-901
5. Mohemmed, A., Schliebs, S., Matsuda, S., Kasabov, N.: SPAN: Spike pattern association neuron for learning spatio-temporal spike patterns. Int. J. Neural Syst. **22**(4), 1–17 (2012). doi:10.1142/S0129065712500128
6. Yu, Q., Tang, H., Tan, K.C., Li, H.: Precise-Spike-Driven synaptic plasticity: learning hetero-association of spatiotemporal spike patterns. PLoS ONE **8**(11), 1–16 (2013). doi:10. 1371/journal.pone.0078318
7. Gardner, B., Grüning, A.: Supervised learning in spiking neural networks for precise temporal encoding. PLoS ONE **11**(8), 1–28 (2016). doi:10.1371/journal.pone.0161335
8. Bondarev, V.N.: Pravila obucheniya impul'snogo nejrona dlya adaptivnoj obrabotki signalov (Training rules of pulse neuron for the adaptive signal processing). In: Proceedings of the XVIII All-Russian scientific and technical conference "Neuroinformatics-2016", part 2, pp. 192–202. NIYaU MIFI Publ., Moscow (2016)
9. Bondarev, V.: Vector-matrix models of pulse neuron for digital signal processing. In: Cheng, L., Liu, Q., Ronzhin, A. (eds.) Advances in Neural Networks—ISNN 2016. Lecture Notes In Computer Science, vol. 9719, pp. 647–656. Springer, Cham (2016). doi:10.1007/978-3-319-40663-3_74
10. Rusu, C.V., Florian, R.V.: A new class of metrics for spike trains. Neural Comput. **26**(2), 306–348 (2014). doi:10.1162/NECO_a_00545
11. Rossum, M.C.W.: Novel Spike Distance. Neural Comput. **13**, 751–763 (2001)
12. Bondarev, V.N., Smetanina, T.I.: Adaptivnyj sintez cifrovogo fil'tra dlya akselerometricheskogo volnografa (Adaptive synthesis of the digital filter for accelerometer wave gage). Sistemy kontrolya okruzhayushchej sredy. **2**(22), 25–28 (2015)

Information Environment for Neural-Network Adaptive Control System

Le Ba Chung[1]([✉]) and Y.A. Holopov[2]

[1] Moscow Institute of Physics and Technology (State University), Dolgoprudny, Russia
chungbaumanvietnam@gmail.com
[2] Lebedev Institute of Precision Mechanics and Computer Engineering, Moscow, Russia

Abstract. The computer architecture for the neural-network control system with a synchronous peripheral subsystem allows transmitting information about the state of the control object and its environment with maximum accuracy. Synchronization between executive resources eliminates restrictions on the interaction between the components of complex control objects. This problem used to be able to be solved only by breaking the problem into slightly dependent smaller tasks. The proposed model is a coherent information environment for neural-network control system, with simultaneous record mode of the object state parameters. This mode is important for the control systems objects with unidentified degrees of freedom – typical field of applications of neural systems. The proposed model of the synchronous information environment is necessary even the control system is implemented with a fixed algorithm.

Keywords: Coherent information environment · Neural network · Digital neural-network control system · Subsystem of input-output

1 Introduction

Like any digital control system (DCS), a digital neural-network control system (DNCS) consists of: a subsystem of measurement – sensors and analog-to-digital converters measure the parameters of the control object (CO) and its surroundings; the calculation subsystem, which calculates and generates control actions; and executive subsystem – digital-to-analog converters and actuators directly to the control the component parts of the CO and interaction with the environment.

The function of the calculation subsystem in DNCS performs the computer implemented on the basis of a neural network. At the beginning of the life cycle of the DNCS, neural network is not trained [1–4], therefore, the control algorithm is missing, the model of the control object is not formed; the degrees of freedom, dynamic parameters, etc. are unknown. The neural network training occurs as a result of simulation and evaluation of control experiments. Every following experiment is performed on the object, formally, with the new DCS, which is generated as a result of previous experiments.

Unlike traditional DCSs with hard control algorithms, in DNCS the information from the subsystem of measurement is used not only to calculate control actions but also for

B. Kryzhanovsky et al. (eds.), *Advances in Neural Computation, Machine Learning, and Cognitive Research*, Studies in Computational Intelligence 736, DOI 10.1007/978-3-319-66604-4_9

correction of models of the control object. It is because of the need to use the state parameters of the control object for the adequate formation of the model of the object to the subsystem of measurement of the DNCS special requirements.

We explain the last statement in more detail. The comparison will hold, and consider the traditional, centralized DNCS as an alternative digital control system after all, the DCS with a neural network computer, due to the uncertainty of the model of the CO, is built as the system with a single central computer.

Traditional digital control systems are implemented as a hierarchical set of regulators, coordinated control the object in the mode of loosely related tasks. Each of these tasks controls one of the degrees of mechanical, electrical, functional, etc. freedom of control object. The structure and intensity of information exchanges in the composite parts like DCS to reflect the degrees of independence of control object parts, in other words, the structure of the DCS follows the structure of the control object, is decomposed by degrees of its freedom. The operation of the object in each of the degrees of freedom are relatively independent, so we can control the object as a whole, through simplified control tasks of parts, which tasks account for limited, local set of state parameters of the CO. This set of state parameters is necessary for this local control task.

We now consider the time profile of the practice control tasks in traditional DCS.

Each of the local control tasks – control can function quite independently from the rest of not only algorithmically, but also in time, i.e. the duration of the control cycles in independent tasks can be different, and the phases of the operation, even if the same duration of the cycle is asynchronous. In centralized, traditional digital control systems with a fixed control algorithm, the local control tasks are processed by the computing subsystem in the split mode of time – sequentially. Measuring sub-task of each local control problem generates a limited set of status settings for that task only. Individual state components are sampled in the close moments of time and the simultaneity of their survey, does not affect the quality of governance in the local control loop. However, in a single computer, phase of the survey parameters in different regulatory tasks, separated in time is already noticeable, but control of the object is possible, because the control is executed through independent, locally sustainable degrees of freedom, with the agreement of the other slower time scale.

Unlike traditional DCS, in DNCS initial structuring of control tasks does not exist, therefore there is no possibility of partitioning the full set of state parameters of the CO and its surrounding on the group phase of formation which should be close in time and may vary from group to group. Therefore, a set of state parameters for DNCS should be formed as a single array simultaneously recorded and measured state variables. Due to the fact that the components of this array relevant at the same time and can only be edited at the same time, it is logical to call the state vector of the control object. The described situation is similar to the motion: periodic discrete recording and playback of continuous processes will be valid if all the objects in the frame and their connections in that frame are recorded at the same time, because the viewer (in DCS – analysis) of all parts of the frame happens in parallel and simultaneously. Continuing the analogy we can say that for DNCS we need to find a way of perception of the parameters of the CO and his associates in the so-called «instant snapshot» in time of the control object. Next, the paper describes the technology of creating information environments that are suitable for simultaneous, coordinated control of subsystem of measurement of the DNCS.

2   Analysis of the Structure of a Digital Neural Network Control System Based on the Universal Computer

A trained digital neural network control system is a hierarchical set of control loops. Moreover, the structure of any loop repeats a common structure of a digital control system. As a rule, tasks: reception of state parameters, calculation, issuing of control signals in all the control loops are addressed in the framework of one cycle of control.

In Fig. 1 shows a typical structure of a digital neural network control system.

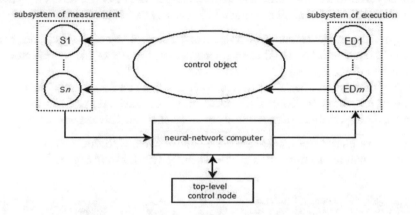

Fig. 1. Generalized structure of the neural network control system

From sensors (S1…Sn), after analog-digital conversion (ADC) into a computer, data on the state of the CO are delivered. Using the received state data, the neural network computer adjusts the model of the control object. On the basis of current models of the CO and target parameters of the action of the CO, the calculator calculates the amount of control and, after digital-to-analog conversion (DAC), sends it to the executive devices - actuators (ED1…EDm), which implement command on a control object.

The computer, in modern DNCS, is implemented in the architecture of the mainframe and operates in asynchronous multitask mode.

Because of the lack of generated control algorithm in untrained DNCS, in this DNCS it is impossible to execute serial calculation of the control loops. Therefore, the calculation in the control cycle is organized in the style of a group of functionally similar operations, starting with the measurement phase for all sensors, and ends with the phase of issuing set-points for all actuators, and all phases of the control cycle are implemented in the software.

To connect sensors and actuators to the central processor of a computer, a set of standard interfaces is used. With rare exceptions, the standard interfaces implemented in the universal microprocessor, the hardware level is supported byte (word) stream transaction and the transaction start is always performed in software. Also, in the time domain, the formation of structural units (state packets and control packets) is implemented programmatically. These packets circulate in interface channels. In microprocessors there is no broadcast mechanism to control the active interfaces, even in a minimal functionality in the form of a simultaneous launch exchange control packs, and software control interfaces of the input subsystem does not enable to implement universal computer strict mode «instant snapshot».

3   Implementation of Phases of the Control Cycle of the Neural Network System

As already mentioned, a specific feature of the algorithm of the DNCS is the presence of the enlarged phase of similar operations in each cycle of operation:

phase of measurement (capturing the state of the control object in the sensors, convert analog parameters into digital code – state vector, transfer of the state vector in the memory of the computer);
phase of calculation (the calculation of control actions on the control object);
phase of execution (delivery the control vector to actuators, perform new control actions, the expectations of the control object reactions, before the next control cycle).

The above phases are repeated in each control cycle with the constant period [5]. The cyclical nature of information processes in DHCS shown in Fig. 2.

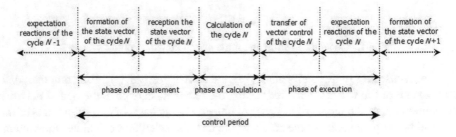

Fig. 2. Phases of the cycle of operation of the digital neural network control system

Peripheral subsystem of modern microprocessors does not allow the control on sensors of the DNCS in the style, which is necessary for the implementation of the measurement phase of the control cycle. Therefore, it is necessary to consider the possibility of hardware support mode «instant snapshot» out of a microprocessor. It is possible to formulate the ability of such coherent subsystem of input data into a computer of the DNCS:

hardware implementation of batch transaction interfaces, the interaction with sensors of the DNCS;
synchronous start-mode of the transaction, which fixes the parameters of the control object;

hardware formation mechanism of a single set of state parameters of the CO – the state vector;

hardware start-mechanism of the state packet transfer into computer.

These functions have as their goal that is the formation of an «instant snapshot» and these functions are implemented in a coherent information environment on the basis of the operations of synthesis and analysis of time intervals. The hardware of the modern programmable logic integrated circuits (FPGA) is ideal for such tasks, and the tasks of the synchronous control of hardware functions with time. The strict temporary performing frames, which perform communication transactions with sensors, enable to implement an «instant snapshot» with maximum precision.

If the same model of peripheral control nodes is implemented in the output subsystem, it will be possible to reach the precise control of the phase of issuing control actions individually for each actuator. This will make it possible to carry out precisely the synchronous work of almost all resources of the control object, for example, mobile platforms and transportable useful load.

4   A Coherent Information Environment Model for Neural-Network Control System

On the basis of the analysis, the developed model is a coherent information environment for neural network control system [6]. Its structural scheme is shown in Fig. 3.

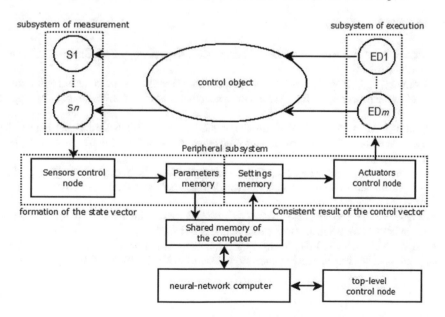

Fig. 3. A coherent information environment for neural network systems

According to this model, all nodes in the system operate at a fixed, pre-established schedule. The lack of asynchrony in the interaction with peripheral devices enables to capture an «instant snapshot» of the control object. Snapshot is a simultaneous fixing of all state parameters of the control object. In a synchronous manner, we undertake all the functions of the input/output and create for DNCS an idealized information environment.

Components of the state vector are actually inputs of the neural network of computer. Regardless of the control algorithm in neural network control system, all components of the state vector entering the processing should be relevant at the same time in the limit of one cycle of system operation. The synchronous architecture of the subsystem input-output allows you to create a parameters package of the control object with the desired DNCS properties.

5 Conclusion

The computer architecture for DNCS with synchronous peripheral subsystem allows maximum accuracy to transmit information about the state of the control object and its environment. Synchronous work of executive resources will enable to remove restrictions on the interaction of subsystems of complex control objects, which are forced to operate in the mode of loosely related tasks.

The proposed model of the synchronous information environment relevant in the implementation of control systems with a fixed algorithm.

References

1. Zhdanov, A.A.: Autonomous artificial intelligence (monograph), p. 358. BINOM, Laboratory Knowledge (2008)
2. Ponomarev, D.Y., Zhdanov, A.A., Chernodub, A.N.: Neural network implementation formal model of the neuron used for "Autonomous adaptive control". Neurocomput. Dev. Appl. 1, 64–75 (2005)
3. Zhdanov, A.A.: On the Method of Autonomous Adaptive Control. Neurocomputers and Their Application. Materials of International Youth Scientific School, pp. 20–67. Publishing House TRTU, Taganrog (2007)
4. Zhdanov, A.A., Preobrazhensky, N.B., Holopov, Y.A., Stepanyan, I.V., Trung, N.H.: Hardware implementation of a neural network in an adaptive control system. Neurocomput. Dev. Appl. 6, 55–62 (2016)
5. Holopov, Y.A., Preobrazhensky, N.B.: Cyclic digital control. International Scientific Institute "Educatio". Monthly Sci. J. Part 3 Tech. Sci. 7, 39–41 (2014)
6. Holopov, Y.A., Chung, L.B., Trung, N.T.: A coherent information environment for highly dynamic control system. Mechatron. Autom. Control. 12(17), 816–820 (2016)

Neural Network Semi-empirical Modeling of the Longitudinal Motion for Maneuverable Aircraft and Identification of Its Aerodynamic Characteristics

Mikhail Egorchev and Yury Tiumentsev$^{(\boxtimes)}$

Moscow Aviation Institute, National Research University, Moscow, Russia
mihail.egorchev@gmail.com, yutium@gmail.com

Abstract. A modelling and simulation problem is considered for longitudinal motion of a maneuverable aircraft that is viewed as a nonlinear controlled dynamical system under multiple and diverse uncertainties. This problem is solved by utilizing semi-empirical neural network based approach that combines theoretical domain-specific knowledge with training tools of artificial neural network field. Semi-empirical approach allows for a substantial accuracy improvement over traditional purely empirical models such as the Nonlinear AutoRegressive neural network with eXogenous inputs (NARX) It also provides solution to system identification problem for aerodynamic characteristics of an aircraft, such as the coefficients of aerodynamic axial and normal forces, as well as the pitch moment coefficient. Representative training data set is obtained using an automatic procedure which synthesizes control actions that provide a sufficiently dense coverage of the region of change in the values of variables describing the simulated system. Neural network model learning efficiency is further improved by the use of special weighting scheme for individual training samples. Obtained simulation results confirm the efficiency of proposed simulation approach.

Keywords: Nonlinear dynamical system · Semi-empirical model · Neural network · System identification

1 Introduction

When designing an aircraft, one of the most important problems is identifying of its aerodynamic characteristics. An approach was proposed in [1–3] to solve this problem using semi-empirical artificial neural network (ANN) models of nonlinear controlled dynamical systems. These semi-empirical models are based on the grey-box model concept introduced in [4,5].

This approach differs significantly from the traditionally accepted method for solving problems of this class, which are based on the use of the linearized model of the aircraft disturbed motion. The conventional approach uses the

© Springer International Publishing AG 2018
B. Kryzhanovsky et al. (eds.), *Advances in Neural Computation, Machine Learning, and Cognitive Research*, Studies in Computational Intelligence 736,
DOI 10.1007/978-3-319-66604-4_10

representation of the dependences for the aerodynamic forces and moments in the form of their Taylor series expansion, leaving in it, as a rule, only members not higher than the first order.

Accordingly, the solution of the identification problem with the conventional approach is reduced to reconstructing from the experimental data the dependences describing the coefficients of the Taylor expansion, in which the derivatives of the dimensionless coefficients of the aerodynamic forces and moments with respect to the various parameters of the motion of the aircraft (C_{L_α}, C_{y_β}, C_{m_α}, C_{m_q} etc.) are determining.

In contrast, the semi-empirical approach performs the reconstruction of the relations for the force coefficients C_D, C_L, C_y and moments C_l, C_n, C_m as some whole non-linear dependences on the corresponding arguments, without their series expansion and linearization, i.e. the functions themselves, represented in the ANN-form, are evaluated, and not the coefficients of their expansion in the series. Each of these dependences is implemented as a separate ANN-module, built into a semi-empirical ANN-model. Derivatives C_{L_α}, C_{y_β}, C_{m_α}, C_{m_q} and others can be found, if necessary, using the results obtained during formation of the ANN-modules for the coefficients of forces and moments within the semi-empirical ANN-model.

A mathematical model of the longitudinal motion of a maneuverable aircraft is derived, which is used as a basis in the formation of the corresponding semi-empirical ANN-model, as well as for the generation of a training set. An algorithm for this generation is proposed, which provides a fairly uniform coverage of the possible values of state variables and controls for the maneuverable aircraft by training examples. Next, a semi-empirical ANN-model of the longitudinal controlled motion of the aircraft is formed, including the ANN-modules realizing the functional dependences for the coefficients C_D, C_L and C_m. The identification problem for these coefficients is solved when learning the obtained ANN-model. The corresponding simulation results characterizing the accuracy of the obtained ANN-model as a whole are given as well as the accuracy of the solution of the identification problem for aerodynamic coefficients.

2   Mathematical Model of Longitudinal Motion for Maneuverable Aircraft

To solve the problem, it is required to form a source mathematical model of the longitudinal motion of an aircraft. This model is represented by a system of nonlinear ordinary differential equations (ODE), traditional for aircraft flight dynamics [6].

The model consists of 9 equations of the first order for aircraft state variables, including 4 equations for variables V_T, γ, R and h, describing trajectory aircraft motion; 2 equations for the variables Θ and q, describing angular aircraft motion; 1 equation for the variable $\bar{T}$ for aircraft engine power level response; 2 equations for variables δ_e and $\dot{\delta}_e$ describing the actuator dynamics for the aircraft elevator. Here V_T is aircraft total velocity, m/s; γ is flight path angle, deg; R is range of flight, m; h is altitude, m; Θ is pitch angle, deg; q is body-axis pitch rate, deg/s; $\bar{T}$ is actual power level of the aircraft engine; δ_e is elevator deflection, deg; $\dot{\delta}_e$ is rate of elevator deflection, deg/s. The right-hand sides of the equations of motion contain the relations for aerodynamic forces, axial C_X and normal C_Z, and also for the aerodynamic pitch moment C_m. These relations are non-linear functions of the appropriate arguments, namely, the angle of attack α and also V_T, δ_e and q. Command signal of the elevator actuator $\delta_{e_{act}}$ and engine throttle setting δ_{th} were used as control signals.

The concretization of this model of motion was carried out for the case of a maneuverable F-16 aircraft. The required data characterizing this aircraft, including the model of its engine, are taken from [7]. The computational experiments performed with this model were carried out in the altitude range from 1000 m to 9000 m and in the range of Mach numbers from 0.1 to 0.6.

3   Generation of a Representative Set of Training Data

When solving problems of the considered type, one of the most important tasks is the generation of a representative set of data that characterizes the behavior of the simulated dynamic system on a rather large range of values for the system state and control variables. This task is critically important for obtaining a authentic dynamic system model, but it has no simple solution. The required training data for the generated ANN-model can be obtained by means of specially organized test excitations for the simulated system.

The training set used in the experiments described in this article was formed using an automatic procedure proposed by the authors. This procedure synthesizes control actions that provide a sufficiently dense coverage of the region of change in the values of variables describing the simulated system. Then, the resulting set of control actions is applied to the simulation object and the obtained trajectories are used to generate the training set. The test set is formed in a similar way.

In addition to the representative training set, we use the weighting of individual examples from the training set to improve the learning efficiency for the ANN-model. It is based on the following considerations. If the arguments of the K examples from the training set are located in a small neighborhood, then this situation is analogous to giving weight K to some average example from this region. Thus, the irregular distribution of examples can lead to increased model accuracy in some areas due to its lowering in others. In order to avoid such a situation, after completing the procedure for synthesizing the training set, the elements of this set are weighed. For this purpose, set of vectors Λ is formed. Vectors $\lambda \in \Lambda$ consist of control variables and state variables of each selected trajectory $\langle u(t), x(t) \rangle \in Q$ at each moment of time $t \in [T_{min}, T_{max}]$, where $u(t)$ and $x(t)$ are control and state vector of simulated dynamical system. For each element $\lambda \in \Lambda$, we search for elements located in its ε-neighborhood. Then, the corresponding example from Q is assigned a weight inversely proportional to the number of neighbors found.

When implementing this algorithm on a computer, you should select an appropriate data structure for representing the set Λ and some auxiliary sets associated with it. This structure should ensure the effective execution of operations for finding the nearest neighbor, searching for neighbors in a given neighborhood, and adding new elements in the generated training set. In this paper, as such a structure, we used a k-dimensional tree, namely, its implementation in the FLANN library [8].

This algorithm was successfully used to generate a training set for a semi-empirical ANN-model of the longitudinal motion for the F-16 maneuverable aircraft. The following range of variables was considered: $\delta_{e_{act}} \in [-25^0, 25^0]$, $\delta_e \in [-25^0, 25^0]$, $\delta_{th} \in [0, 1]$, $\bar{T} \in [0, 100]\%$, $\gamma \in [-90^0, 90^0]$, $q \in [-100, 100]$ deg/s, $V_T \in [35, 180]$ m/s, $\alpha \in [-20^0, 90^0]$.

4   Semi-empirical Neural Network Model of Aircraft Longitudinal Motion

A general approach to the formation of semi-empirical ANN-models of controllable dynamical systems was presented in [1,2]. For the problems of identification of aircraft aerodynamic characteristics these models are considered in [3], where using the mathematical model of complete aircraft angular motion were solved the problem of finding relationships for aerodynamic lateral and normal force coefficients C_Y and C_Z as well as for aerodynamic rolling, pitching and yawing moment coefficients C_l, C_m and C_n. In this section, we build a semi-empirical ANN-model of the aircraft longitudinal motion, based on the mathematical model mentioned above. This ANN-model allows us to find the relations for the coefficients C_X, C_Z and C_m, with respect to the vast range of possible values of the variables on which these relations depend.

The training and test sets were formed according to the procedure described in the previous section, with a sampling step $\Delta t = 0.01$ s. The vector of state variables is partially observable $y(t) = [V_T(t), \alpha(t), q(t)]^T$, $\alpha = \Theta - \gamma$. The output of the system is affected by additive white noise with root mean square deviation (RMS) $\sigma_{V_T} = 0.01$ m/s, $\sigma_\alpha = 0.01$ deg, $\sigma_q = 0.005$ deg/s.

Training of semi-empirical ANN-models is a non-trivial task. The appropriate algorithms for solving it are considered in [3]. This training is carried out in the Matlab system for neural networks in the form of LDDN (Layered Digital Dynamic Networks) using the Levenberg-Marquardt optimization algorithm based on the root-mean-square error of the model [9]. The Jacobi matrix is calculated using the RTRL (Real-Time Recurrent Learning) algorithm [10].

ANN-modules for nonlinear functions C_X, C_Z and C_m are formed as sigmodal feed-forward networks. As inputs of each of the modules, the values of α, δ_e and q/V_T are taken. The ANN-modules for the C_X and C_Z functions have two hidden layers, the first of which includes 10 neurons and the second one contains 20. The ANN-module for the C_m function has three hidden layers, the first of which includes 10 neurons, the second one has 15 and the third has 20 neurons.

The simulation error on the test set for the obtained semi-empirical ANN-model of the longitudinal motion for the maneuverable aircraft is: $\text{RMS}_{V_T} = 0.00026$ m/s, $\text{RMS}_\alpha = 0.183$ deg, $\text{RMS}_q = 0.0071$ deg/s.

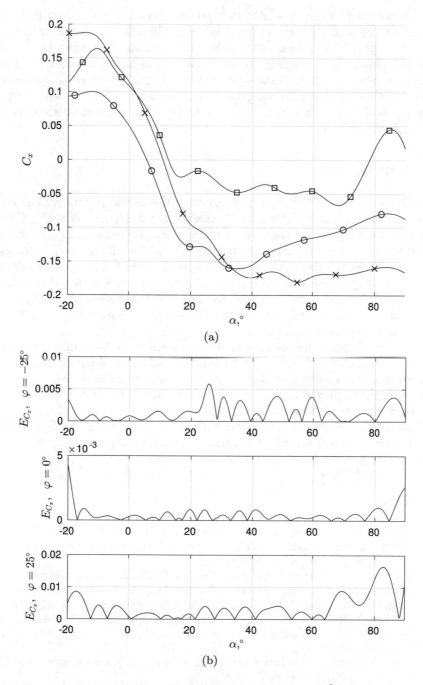

Fig. 1. Simulation results: **(a)** – coefficient $C_X(\alpha, \delta_e)$ for $\delta_e = -25^0$ (marker $\square$), $\delta_e = 0^0$ (marker $\circ$) and $\delta_e = -25^0$ (marker $\times$) according to [7]; **(b)** – approximation error E_{C_X} for fixed values of $q = 0$ deg/s and $V_T = 150$ m/s

The accuracy of the dependences representation for the aerodynamic coefficients can be seen from the example of the coefficient C_X as shown in Fig. 1. The upper part of this figure shows the actual values (according to the data from [7]) of the C_X depending on the angle of attack and the elevator deflection angle. The lower part of the figure shows the errors with which appropriate ANN-module reproduces the corresponding dependence. It can be seen that the accuracy achieved is very high. The results for the other two coefficients (C_Z and C_m) look similar.

5 Conclusions

The results presented above allow us to draw the following conclusions. As in the case described in [3] for the coefficients of aerodynamic forces C_L, C_y and moments C_l, C_n, C_m, methods of semi-empirical ANN-modeling provide the possibility to solve successfully the problem of longitudinal force coefficient identification if the characteristics of the engine are known. If data for these characteristics are not available, then the result of solving the identification problem will be the relationship for the total coefficient of axial force, whose arguments will include the δ_{th} control variable. Usually this is quite enough to simulate the motion of the aircraft.

The second important conclusion, which follows from the obtained results, is that the "computational power" of the semi-empirical ANN-model is quite sufficient to represent complex nonlinear functional dependencies defined on a broad range of their argument values, provided that there is a training set possessing the required level of representativeness.

The simulation results demonstrate the high accuracy of both the ANN-model of the obtained aircraft longitudinal motion and high representation accuracy for corresponding aerodynamic characteristics.

References

1. Egorchev, M.V., Kozlov, D.S., Tiumentsev, Y.V., Chernyshev, A.V.: Neural network based semi-empirical models for controlled dynamical systems. J. Comput. Inf. Technol. **9**, 3–10 (2013). (in Russian)
2. Egorchev, M.V., Kozlov, D.S., Tiumentsev, Y.V.: Neural network adaptive semi-empirical models for aircraft controlled motion. In: Proceedings of the 29th Congress of the International Council of the Aeronautical Sciences, vol. 4 (2014)
3. Egorchev, M.V., Tiumentsev, Y.V.: Learning of semi-empirical neural network model of aircraft three-axis rotational motion. Optical Memory Neural Netw. **24**(3), 201–208 (2015)
4. Oussar, Y., Dreyfus, G.: How to be a gray box: dynamic semi-physical modeling. Neural Netw. **14**(9), 1161–1172 (2001)
5. Dreyfus, G.: Neural Networks – Methodology and Applications. Springer (2005)
6. Bochkariov, A.F., Andreyevsky, V.V., Belokonov, V.M., Klimov, V.I., Turapin, V.M.: Aeromechanics of Airplane: Flight Dynamics, 2nd edn. Mashinostroyeniye, Moscow (1985). (in Russian)

7. Nguyen, L.T., Ogburn, M.E., Gilbert, W.P., Kibler, K.S., Brown, P.W., Deal, P.L.: Simulator study of stall/post-stall characteristics of a fighter airplane with relaxed longitudinal static stability. Technical report TP-1538, NASA, December 1979
8. Muja, M., Lowe, D.G.: Scalable nearest neighbor algorithms for high dimensional data. IEEE Trans. Pattern Anal. Mach. Intell. **36**(11), 2227–2240 (2014)
9. Haykin, S.: Neural Networks: A Comprehensive Foundation, 2nd edn. Prentice Hall PTR, Upper Saddle River (1998)
10. Jesus, O.D., Hagan, M.T.: Backpropagation algorithms for a broad class of dynamic networks. IEEE Trans. Neural Netw. **18**(1), 14–27 (2007)

Dump Truck Fault's Short-Term Forecasting Based on the Multi-agent Adaptive Fuzzy Neuronet

Ekaterina A. Engel[(✉)]

Katanov State University of Khakassia, Lenina 90, Abakan 655017, Russia
Ekaterina.en@gmail.com

Abstract. Dump truck fault's short-term forecasting is the important step for solving real-time fleet dispatching tasks and to provide reliable, efficient and safe open-pit mining. This paper presents a multi-agent adaptive fuzzy neuronet for dump truck fault's short-term forecasts. The agents of the multi-agent adaptive fuzzy neuronet are fulfilled based on recurrent networks. An automatic determination of the optimal architecture's parameters of a neuronet is the most critical task. In order to train the effective multi-agent adaptive fuzzy neuronet we use algorithm, in which the multi-dimensional Particle Swarm Optimization is combined with the Levenberg-Marquardt algorithm. The multi-dimensional Particle Swarm Optimization is first applied to globally optimize the multi-agent adaptive fuzzy neuronet's structure, and then Levenberg-Marquardt is used to speed up convergence process. The simulation results show that proposed training algorithm outperforms multi-dimensional Particle Swarm Optimization and Levenberg-Marquardt algorithm in training the effective multi-agent adaptive fuzzy neuronet for dump truck fault's short-term forecasts.

Keywords: Multi-agent adaptive fuzzy neuronet · Multi-dimensional Particle Swarm Optimization · Dump truck fault's short-term forecasting

1 Introduction

Nowdays, SUEK is one of the largest coal companies in the world and the number one coal producer in Russia. It attempts to find intelligent technologies to improve the key areas of their operations. Dump trucks are equipped with sensors and communication devices. Dump truck fault's short-term forecasting is the important step for solving real-time fleet dispatching tasks and to provide reliable, efficient and safe open-pit mining. For more efficient dump truck fault's short-term forecasting, it is important to take uncertainties into account. These uncertainties originate from dump truck's punishment or under random perturbations of weather, which cause complex dynamics of dump truck fault's time series. Several algorithms have been developed to overcome the given difficulties.

B. Kryzhanovsky et al. (eds.), *Advances in Neural Computation, Machine Learning, and Cognitive Research*, Studies in Computational Intelligence 736,
DOI 10.1007/978-3-319-66604-4_11

The most important approaches are those that provide effective data processing based on intelligent algorithms. The aforementioned approaches include combining evolutionary intelligent agents, neuronets and fuzzy logic. This paper presents a multi-agent adaptive fuzzy neuronet (MAFN) for dump truck fault's short-term forecasts. The results of the MAFN on the challenging real-world problems [1] revealed its following advantages: supports the real time mode and competitive performance, as compared to classical methods; trained MAFN effectively processes the noisy data. The agents of the MAFN are fulfilled based on recurrent networks. An automatic determination of the optimal architecture's parameters of a neuronet is the most critical task. The effective network architecture is up-to-date designed by a human expert, requiring an extensive analysis of the system and the trial-error process. This process is difficult because it is hard to anticipate all conditions of optimal neuronet architecture. The global optimum provided by the multi-dimensional Particle Swarm Optimization (MD PSO) [2] process corresponds to an optimum multi-agent adaptive fuzzy neuronet architecture where the MAFN architecture's parameters (delays, a number of nodes in hidden layer, corresponded weights and biases) are generated from the global optimum. Furthermore, the MD PSO provides a ranked list of MAFN configurations, from the best to the worst. This is an important information, arguing which configurations can effective solve a particular problem. In order to train the effective multi-agent adaptive fuzzy neuronet we use algorithm, in which the MD PSO is combined with the Levenberg-Marquardt algorithm [3]. The MD PSO is first applied to globally optimize the network's structure, and then Levenberg-Marquardt is used to speed up convergence process. The simulation results show that proposed training algorithm outperforms MD PSO and Levenberg-Marquardt algorithm in training the effective MAFN for dump truck fault's short-term forecasts. The multi-agent adaptive fuzzy neuronet was fulfilled based on extensive empirical database, collected from an open-pit mine.

2 The Multi-agent Adaptive Fuzzy Neuronet for Dump Truck Fault's Short-Term Forecasts

Sources of database's information are position, speed, course (GPS) and gar load sensors. The multi-agent adaptive fuzzy neuronet is trained based on the "KARIER" database of aforementioned dump truck's sensors. This database was collected at the Chernogorsky open pit mine from 2013 till 2016. On the basis of dispatching reports moments of dump truck's malfunction during the period up to 10 days is defined. All possible dump truck's malfunctions were divided into 3 classes. The dump truck's malfunction classes P_h^t are qualitative and were coded as appropriate, where $h = \overline{1..32}, t = \overline{1..1461}$. The aforementioned classification includes the following classes:

- class 1: the dump truck's malfunction will have occurred within the next 2 days, $P_h^t = (1,0)$;
- class 2: the dump truck's malfunction will have occurred within period from 3 to 10 days, $P_h^t = (0,1)$;

- class 3: the dump truck's malfunction will have not occurred within the next 10 days, $P_h^t = (0,0)$.

The multi-agent adaptive fuzzy neuronet is fulfilled based on the data

$$s_h^t = (x_h^t = (V_h^t, Vm_h^t, W_h^t, Wm_h^t, I_h^t, Im_h^t, Tm_h^t), P_h^t), \qquad (1)$$

where P_h^t is dump truck's malfunction class; v^t is speed of a dump truck, vm is an maximum speed limit of a dump truck (km/h), d^t is a time period of corresponded condition from last dump truck's malfunction, $V_h^t = \sum_{v^t \leq vm}(v^t d^t)$, $Vm_h^t = \sum_{v^t > vm}(v^t d^t)$, $Tm_h^t = \sum_{T^t > tm}(T^t d^t)$, T^t is an ambient temperature; p^t is the pressure of a dump truck's tire (an indicator corresponded to the average tonnages of the transported mined minerals), pm is a maximum pressure limit of a dump truck, $W_h^t = \sum_{p^t \leq pm}(p^t d^t)$, $Wm_h^t = \sum_{p^t > pm}(p^t d^t)$; i^t is the indication of an inclinometer (degrees), im^t is a maximum indication of an inclinometer limit of a dump truck, $I_h^t = \sum_{i^t \leq im}(i^t d^t)$, $Im_h^t = \sum_{i^t > im}(i^t d^t)$. The number of samples is 46752 ($h * t = 46752$).

2.1   The Training Algorithms of the Multi-agent Adaptive Fuzzy Neuronet

The multi-agent adaptive fuzzy neuronet architecture's parameters (delays, number of nodes in hidden layer, corresponded weights - w and biases) have been coded into particles a. The agents $g_{jq}(x_h^t, w_{jq})$ of the multi-agent adaptive fuzzy neuronet $f(x_h^t, a)$ are fulfilled as two-layered recurrent networks. In this research an agent's number is three. The fitness function evaluated as follows:

$$E(f) = (1/46752) \sum_{h=1, t=1}^{32,1461} \left\| P_h^t - f(x_h^t, a) \right\|. \qquad (2)$$

In order to train the multi-agent adaptive fuzzy neuronet f we use MD PSO [2]. With the encoding of the multi-agent adaptive fuzzy neuronet structure into particles, MD PSO provides not only the positional optimum in the error space, but as well the optimum dimension of space of a task and the dimensional optimum in the neuronet structure space. The dimension range of the MD PSO is $D_{min} = 43, D_{max} = 313$. The MD PSO method includes three steps:

Step 1. The MD PSO (termination criteria: $\{IterNo, \varepsilon_c, ...\}$)
For $\forall a \in 1, S$, do Randomize $xd_a(0), vd_a(0)$ Initialize $\widetilde{xd}_a(0) = xd_a(0)$
For $\forall d \in D_{min}, D_{max}$ do Randomize $xx_a^d(0), xv_a^d(0)$ Initialize $xy_a^d(0) = xx_a^d(0)$
End For. End For.
Step 2. For $\forall i \in \{1, IterNo\}$ Do: For $\forall a \in \{1, S\}$ Do:
If $(f(xx_a^{xd_a(i)})) < min(f(xy_a^{xd_a(i)}(i-1)), min_{p \in S-\{a\}}(f(xx_p^{xd_a(i)}(i))))$
then $xy_a^{xd_a(i)}(i) = xx_a^{xd_a(i)}(i)$.
If $f(xx_a^{xd_a(i)}(i)) < f(xy_{gbest(xd_a(i))}^{xd_a(i)}(i-1))$ then $gbest(xd_a(i)) = a$, where

$gbest(d)$ is global best particle index in dimension d, $xy_{aj}^{xd_a(i)}(i-1)$ is j-th component of the personal best (*pbest*) position of particle a, in dimension $xd_a(i)$.

If $xx_a^{xd_a(i)}(i) < f(xy_a^{\widetilde{xd_a}(i-1)}(i-1))$ then $\widetilde{xd_a}(i) = xd_a(i)$ end If.

If $xx_a^{xd_a(i)}(i) < f(x\overline{y}_{dbest}(i-1))$ then $dbest = xd_a(i)$. End If. End For.

If the termination criteria are met, then Stop.

Step 3. For $\forall a \in 1, S$ Do: For $\forall j \in \{1, xd_a(i)\}$ Do: Compute
$vx_{a,j}^{xd_a(i)}(i+1) = w(i)vx_{a,j}^{xd_a(i)}(i) + c_1 r_{1,j}(i)(xy_{a,j}^{xd_a(i)}(i) - xx_{a,j}^{xd_a(i)}(i)) +$
$c_2 r_{2,j}(i)(x\overline{y}_j^{xd_a(i)}(i) - xx_{a,j}^{xd_a(i)}(i)),$

If $vx_{a,j}^{xd_a(i+1)}(i) \in [X_{min}, X_{max}]$ then $xx_{a,j}^{xd_a(i)}(i+1) = xx_{a,j}^{xd_a(i)}(i) +$
$vx_{a,j}^{xd_a(i)}(i+1)$ else $xx_{a,j}^{xd_a(i)}(i+1) = U(X_{min}, X_{max})$ end If. If
$xx_{a,j}^{xd_a(i+1)}(i+1) \in [X_{min}, X_{max}]$ then $xx_{a,j}^{xd_a(i)}(i+1) = xx_{a,j}^{xd_a(i)}(i+1)$
else $xx_{a,j}^{xd_a(i)}(i+1) = U(X_{min}, X_{max})$

end If. End For. $[v_{min}, v_{max}]$ is dimensional velocity range.

$vd_a(i+1) = vd_a(i) + c_1 r_1(i)(x\widetilde{d}_a(i) - xd_a(i)) + c_2 r_2(i)(dbest - xd_a(i)).$

If $vd_a(i+1) < v_{min}$ then $z = v_{min}$ end If. If $vd_a(i+1) > v_{max}$ then $z = v_{max}$ end If. If $v_{min} \le vd_a(i+1) \le v_{max}$ then $xd_a(i+1) = xd_a(i) + vd_a(i+1)$ else $xd_a(i+1) = xd_a(i) + z$ end If.

If $(xd_a(i+1) < D_{min})$ or $(xd_a(i+1) > D_{max})$ or $(P_d(i) \ge max(15, xd_a(i+1)))$ then $xd_a(i+1) = xd_a(i)$ end If. End For. End For.

The MD PSO provides an optimum multi-agent adaptive fuzzy neuronet architecture $f1$ and corresponded two-layered recurrent neuronetworks architectures g_i. The Levenberg-Marquardt algorithm [3] can be described as follows:

Step 1. Initialize the weights and parameter μ (in this research $\mu = 0.01$).

Step 2. Compute the train error $E(f1)$ as (2).

Step 3. Due to obtain the increment of weights Δw_i we solve the following equation

$$\Delta w_i = [J_i^T J_i + \mu x]^{-1} J_i^T E(f1),$$

where J_i is the Jacobian matrix, μ is the learning rate which is to be updated using the β depending on the outcome.

Step 4. Update $w_i = w_i + \Delta w_i$. Recomputed the train error E'(f) according (2).

Step 5. If $E'(f1) > E(f1)$ then $w_i = w_i + \Delta w_i$; $\mu = \mu\beta$; Go to step 2 else $\mu = \mu/\beta$; go to step 4 end If.

In order to train the effective agents of the multi-agent adaptive fuzzy neuronet for dump truck fault's short-term forecasts the MD PSO and the Levenberg-Marquardt algorithm have been combined. The MD PSO is first applied to globally optimize the network's structure (the PSO will stop after a global solution is localized within small region), and then Levenberg-Marquardt is used to speed up convergence process.

2.2   The Multi-agent Adaptive Fuzzy Neuronet

The algorithm of the agent's interaction uses a fuzzy-possibilistic algorithm [4] and includes four steps:

Step 1. for each $agent_q$ in subculture S_j do $g_{jq}(x_h^t) \longrightarrow GetResponse(agent_q;$ $x_h^t)$, $\nu_q \longrightarrow TakeAction(g_{jq}(x_h^t))$: Evaluate error as $E(g)(2)$, Calculate $\nu_q = 1 - E$.
End For. $w = [g_{j1}(x_h^t), ..., g_{jq}(x_h^t)]$.
Step 2. Solve equation $[\Pi_{i=1}^q(1 + \lambda w_i) - 1]/\lambda = 1, -1 < \lambda < \infty$.
Step 3. Calculate $s = \int h \circ W_\lambda = sup_{\alpha \in [0,1]} min\{\alpha, W_\lambda(F_\alpha(\nu_j))\}$, where $F_\alpha(\nu_j) = \{F_i|F_i > \alpha\}, \nu_j \in V, W_\lambda(F_\alpha(\nu_j)) = [\Pi_{F_i \in F_\alpha(\nu_j)}^k(1+\lambda w_i)-1]/\lambda$
Step 4. Calculate $f = Fes(g_{jq}(x_h^t)) = max_{\nu_j \in V} s(w_j)$.

The fuzzy-possibilistic method allows for the forecasting of the dump truck's malfunction class in a flexible manner, so as to take into account the responses of all agents based on fuzzy measures and the fuzzy integral.

Fulfillment of the multi-agent adaptive fuzzy neuronet briefly can be described as follows:

Step 1. All samples $N = 46752$ ($N = h * t = 46752$) were classified into two groups according truck's operating mode: A_1 is normal mode ($C_h^t = -1$), A_2 is punishment ($C_h^t = 1$). This classification generates vector with elements C_h^t.
Step 2. Two-layer recurrent network (number of hidden neurons and delays are 7 and 2, respectively): $F(s_h^t)$ was trained. The vector (1) was network's input. The vector C_h^t was network's target. Fuzzy sets A_j (A_1 – normal mode, A_2 – punishment) with membership function $\mu_j(s)$ are formed base on aforementioned two-layer recurrent network $F(s_h^t)$, where $j = \overline{1..2}$.
Step 3. We train based on o optimization algorithm (If $o = 1$ then optimization algorith is MD PSO; If $o = 2$ then optimization algorith is Levenberg-Marquardt algorithm; If $o = 3$ then optimization algorith is proposed algorithm, in which the MD PSO is combined with the Levenberg-Marquardt algorithm) three two-layered recurrent neural networks: $Ir_{jq} = g_{jq}(x_h^t)$ ($j = \overline{1..2}, h = \overline{1..32}, t = \overline{1..1461}$) based on the data (1). The MD PSO provides the optimum number of hidden neurons and delays of aforementioned two-layer recurrent networks. The fitness function is (2). This step provides recurrent neural networks $g_{jq}(x_h^t)$ which forecast the malfunction class of a dump truck. Two agent's subcultures S_j are formed base on aforementioned two-layer recurrent networks.
Step 4. If-then rules are defined as:

$$\Pi_j : If \ s_h^t \ is \ A_j \ then \ f = Fes(g_{jq}(x_h^t)). \tag{3}$$

Simulation of the multi-agent adaptive fuzzy neuronet briefly can be described as follows.

Step 1. For $\forall c \in [1096, 1461] \forall h \in [1, 32]$

Aggregation antecedents of the rules (3) maps input data s_h^c into their membership functions and matches data with conditions of rules. These mappings are then activates the k rule, which indicates the k truck's operating mode and k agent's subcultures – S_k, $k = \overline{1..2}$.

Step 2. According the k truck's operating mode the multiagentny adaptive fuzzy neuronet (trained base on the data s_h^d, where $d = \overline{1..c-1}$) creates the forecasting dump truck's malfunction class $f = Fes(g_{jq}(x_h^c))$ as a result of multi-agent interaction of subculture S_k.

3 Results

To illustrate the benefits of the MAFN in dump truck fault's short-term forecasting, the numerical examples from the previous section are revisited using author's software [5,6]. There the three MAFN were fulfilled based on the data (1). First, the multi-agent adaptive fuzzy neuronet (MAFN1) was trained using MD PSO. Due to obtain statistical results, we perform 120 MD PSO runs with following parameters: $S = 250$ (we use 250 particles), $E = 150$ (we terminate at the end of 150 epochs). Figure 1(a) shows that only two distinct sets of MAFN architecture with dbest = 103 and dbest = 163 can achieve test classification rate

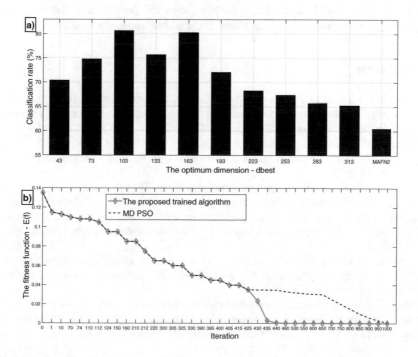

Fig. 1. (a) The MAFN's classification rates (b) The mean convergence curves

above 80,2 over data (1), $t = \overline{1..1461}$. We have chosen simpler solution with dbest=103 as an optimum multi-agent adaptive fuzzy neuronet MAFN1.

Because the MD PSO has a superior generalization ability and a native feature of having a better and faster convergence to an optimum solution in low dimensions, it provides simple but effective and robust MAFN. The MAFN1 has three agents of each subculture S_k. The aforementioned agents are the two-layered recurrent neural network. The agent's number of hidden neurons and delays are 3 and 2, respectively. MAFN2 has same architecture. The second multi-agent adaptive fuzzy neuronet (MAFN2) was trained by Levenberg-Marquardt algorithm. The third multi-agent adaptive fuzzy neuronet (MAFN3) was trained by the proposed algorithm, in which the MD PSO is combined with the Levenberg-Marquardt algorithm. The MD PSO is first applied to globally optimize the MAFN's structure, and then Levenberg-Marquardt is used to speed up a convergence process. Figure 1(b) shows mean convergence curves of the MD PSO and the proposed algorithm for train MAFN. In Fig. 1(a), the ineffectiveness of the MAFN2 can be seen. Figure 1(b) shows that MAFN3 has definitely more convergence speed than MAFN1 in dump truck fault's short-term forecasting. Simulation comparison results for a dump truck's malfunction class short-term forecasting demonstrates the effectiveness of the multi-agent adaptive fuzzy neuronet trained by proposed algorithm, in which the multi-dimensional Particle Swarm Optimization is combined with the Levenberg-Marquardt algorithm as compared with the same ones trained by MD PSO or Levenberg-Marquardt algorithm. The analysis of the evolving errors shows the potential of the multi-agent adaptive fuzzy neuronet in Mine Fleet fault's short-term forecasts.

References

1. Engel, E.A.: Sizing of a photovoltaic system with battery on the basis of the multi-agent adaptive fuzzy neuronet. In: 2016 International Conference on Engineering and Telecommunication (EnT), pp. 49–54 (2016)
2. Kiranyaz, S., Ince, T., Yildirim, A., Gabbouj, M.: Evolutionary artificial neural networks by multi-dimensional particle swarm optimization. Neural Netw. **22**(10), 1448–1462 (2009)
3. Horst, R., Tuy, H.: Global Optimization. Springer, Berlin (1996)
4. Engel, E.A.: The method of constructing an effective system of information processing based on fuzzy-possibilistic algorithm. In: Proceedings of the 15th International Conference on Artificial Neural Networks, Neuroinformatics 2013, part 3, pp. 107–117 (2013)
5. Engel, E.A.: The multi-agent adaptive fuzzy neuronet. Certificate about State registration of computer programs. No. 2016662951, M.: Federal Service for Intellectual Property (Rospatent) (2016)
6. Engel, E.A.: The intellectual system for forecasting of a non-linear technical object's state. Certificate about State registration of computer programs. No 2016663468, M.: Federal Service for Intellectual Property (Rospatent) (2016)

Object Detection on Images in Docking Tasks Using Deep Neural Networks

Ivan Fomin[✉], Dmitrii Gromoshinskii, and Aleksandr Bakhshiev

The Russian State Scientific Center for Robotics and Technical Cybernetics (RTC),
Tikhoretsky Prospect, 21, 194064 Saint-Petersburg, Russia
{i.fomin,d.gromoshinskii,alexab}@rtc.ru

Abstract. In process of docking of automated apparatus there is a problem of determining of them relative position. This problem may be effectively solved with algorithms for relative position calculation, based on television picture formed by camera, installed on one apparatus and observing another one, or docking position. Apparatus position and orientation calculates using visual landmarks positions and information about 3D configuration of observing object and visual landmarks' relative positions. Visual landmarks detection algorithm is the crucial part of such solution. Study of ability of application of object detection system based on deep convolutional neural network to task of visual landmark detection will be discussed in this article. As an example, detection of visual landmarks on space docking images will be discussed. Neural network based detection system learned using images of International Space Station received in process of docking of cargo spacecrafts will be represented.

Keywords: Object detection · Deep neural networks · Convolutional neural networks · Faster R-CNN · Machine learning · Computer vision

1 Introduction

1.1 Relevance of the Problem

One of the most sophisticated and relevant problems in area of automated apparatus docking process is determination of relative position between one apparatus and another, or apparatus and docking position. If both apparatus, or at least one of them, equipped with video cameras, this problem can be solved using positions of visual landmarks in images from the camera. As an example, docking between spacecraft and International Space Station (ISS) will be discussed. Nowadays in process of docking this problem solving with special radio-electronic and optical systems, components of this systems must be placed on ISS and spacecraft. Also, all spacecrafts during last 40 years equipped with specialized television system that using in process of docking for additional visual control.

Earlier in articles [1, 2] described application of television system to determining relative position of spacecraft relative to ISS. Special computer applications developed

© Springer International Publishing AG 2018
B. Kryzhanovsky et al. (eds.), *Advances in Neural Computation, Machine Learning, and Cognitive Research*, Studies in Computational Intelligence 736,
DOI 10.1007/978-3-319-66604-4_12

to solve this task. These applications can be installed on special laptop PC on the ISS or on desktop in Mission Control Center. Applications receive video signal from camera, installed on spacecraft, in process of docking, and performs simultaneous detection and tracking of visual landmarks that exists onboard ISS. Using data of such landmarks' positions, known model of camera, relative positions of landmarks determined by precise 3D model of ISS we able to precisely calculate relative positions of spacecraft and ISS by solving PnP problem.

1.2   Statement of the Problem

One of the most important components of developed system, that determines performance of the system and precision of relative position, determined by system is the module for simultaneous detection and tracking of visual landmarks. Other methods are fully mathematically described and their numerical result fully rely on precision of landmarks' pixel positions determined by this module.

Current television system has some specificities. All components on the way from cameras to PC where our system installed are analog, including components for radio signal transmitting and receiving on the ISS and Control Center. Each of these components have some different negative influence on the signal, and all of them may cause different distortions. Examples of distorted images received in the process of docking represented on Fig. 1 [3].

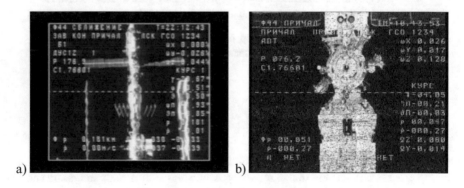

Fig. 1. Examples of distortions: (a) image size distortion, (b) camera matrix noise

On the other side, instead current approach, when to compensate image distortions we use algorithms of different complexity, we can use absolutely different approach. Neural networks at all and partially convolutional neural networks have very good ability to generalization of input learning information. Two years ago, object detection system based on deep convolutional neural network [4] have been introduced. This system utilizes combination of neural networks to detect different objects on images and results with mean average precision up to 78.8%.

We decided to try to apply this system in our problem to detect visual landmarks in images from docking video records, study results of this system in our task and decide, is it make any sense to use this system for object detection in space docking images or other images in our future works.

2 Description of Chosen Systems

2.1 Structure of the Faster R-CNN

To perform our studies, we decided to utilize ready-to-use realization of Faster R-CNN detection system [4] based on the neural networks that implemented using Caffe system and its Python language bindings, named py-faster-rcnn [5].

In this section, we will briefly discuss overall structure of Faster R-CNN and basic principles of how it work. Simplified scheme of the system represented on Fig. 2. Input of the system is the image. Firstly, input image processing by convolutional naurel network. This part of network contains convolutional layers with ReLU error function and pooling layers between them.

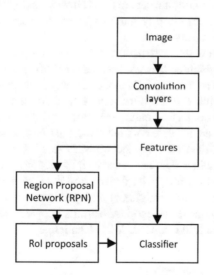

Fig. 2. Scheme of Faster R-CNN system

Convolutional layers perform operation of convolution of small kernels (usually square) with all channels, that passed to the input of the layer. Pooling layers usually choose one better output from each area in each output of previous layer, usually pooling performed in small squares like 2×2 or 3×3.

Outputs of convolutional neural network are feature maps, that firstly goes to the input of special network for generation propositions of regions of interest (RoI), that may contain or not contain some objects.

This network uses results of the convolutional layers to predict possible positions of objects in the source image and possibility of being an object for each such region.

After moment all RoIs are generated, they pass to the input of the classifier network part. Every RoI projects to the output of the last convolutional layer and resulting patch transforms to the vector of standard size and passes to the input of the classifier that is fully-connected neural network.

2.2 Learning of Faster R-CNN

Faster R-CNN standard way of learning is sequential learning of each part of the network. Because networks for RoI proposition and object classification share same convolutional layers in lower part of whole network this convolutional layer can be learned together. Classifier (fully-connected part) waits for fixed region proposals during process of learning, more precisely for RoIs that formed with similar rules for each image in each learning batch. While RPN learning, all weights are changing, rules of RoI selection are also changing and it is very hard for learning process to converge. Then in practice standard learning procedure contains few steps, where RPN and classifier learns sequentially. First, system leans RPN layers with convolutional layers from scratch. Then learned on previous step convolutional layer and RPN using to learn classifier. On third step weights of convolutional net are fixed already, and system performs RPN fine-tuning. And finally, on fourth step system using fixed fine-tuned RPN to finetune fully-connected classification layers.

In accordance with authors' instruction, before learning classifier and convolutional layer initializing with weight values pre-learned on 1000-classes ImageNet dataset and using this weight values in learning steps. Convolutional neural networks are "deep learning" nets, their weights forms in the process of learning from random values, each layer learn convolution kernels or feature detectors of different scale. To learn network to generate very good feature detectors network must be learned on very big count of examples. Because free datasets for object are very small if we compare them to object classification datasets authors decided to use weights of convolutional and classification parts previously learned on ImageNet dataset and for each new category set and last layer config they only fine-tune weights of all layer in assumption that new categories somehow similar to ImageNet categories. Authors showed that this approach significantly improve their result in PASCAL VOC Challenge.

3 Experimental Researches

For detection system learning we prepared dataset with images that received in process of docking of space apparatus that docked in different time to different docking nodes in International Space Station (ISS). Positions of all docking parts visible in each frame were marked by hands.

To improve quality of object detection we tried to apply augmentation to the data. Source list of images was increased five times, including original image and image with height and width increased by 5% and 10% respectively.

Py-faster-rcnn contains three different classification network models, RPN part is fixed for each model. First model is ZF (network from Zeiler-Fergus article), second is VGG-CMM-M-1024, same as ZF but parameters of some layers are modified. Final architecture is VGG-16, that showed best result in PASCAL VOC Challenge in detection discipline, in time when original article was published. All models are described properly in work [6] and deploy versions represented in BVLC Caffe Model Zoo [7].

Unfortunately, VGG-16 architecture not available to experiment because it needs more than 4 Gb of video memory to perform learning. Models ZF and VGG-CNN-M-1024 was both tested on different versions of our datasets, and will be compared in the tables below (Table 1).

Table 1. United table of the results

	Node 1	Node 2	Node 3	Node 4
VGG-1024, all objects, no changes	0,790	0,363	0,309	0,394
VGG-1024, augmented	0,765	0,345	0,297	0,508
VGG-1024, high contrast	0,779	0,377	0,250	0,454
ZF, all objects, no changes	0,708	0,316	0,219	0,345
ZF, high contrast	0,727	0,365	0,217	0,36

We learned both network using three different datasets. First dataset is source set, where all objects from 4 docking nodes are collected in one dataset. Second set is source, that extended using data augmentation, described before. Images in third dataset are equals to first, but each image additionally processed by contrast filter, because all images in source set have low contrast, some of them extremely low.

To receive some comparable quality metrics for our results we decided to use Mean Average Precision quality metrics, that was calculated according to rules, that can be found in full PASCAL VOC Challenge rules document [8]. This metric for now is standard to quality measurement in object detection systems. To make our results short, Precision-Recall curves and mAP calculated for all objects in test set together.

Tests shown us that increasing of learning set multiple times has very low positive effect on detection quality. Contrast increasing has low positive effect for one separate object type, and no effect for the rest objects. Most of best results shown by source network learned without filtering and/or augmentation.

If we compare results to similar with another architect, object detection mAPs with VGG-CNN-M-1024 detection network are a bit higher than respective values with ZF network. Some objects are detected better, another ones – worse.

4   Conclusions

In the process of our research, we performed different experiments of possibility pf application of Faster R-CNN neural-based object detection system to the task of object detection in process of docking, as an example we used space docking images. Received results show us, that system is able to detect objects with good contrast to surrounding frame area, such as special docking target (used by operator to lead spacecraft to the docking node) or large docking unit. Visible features, that has low difference from surroundings, or any other part of the station are almost never detected. Augmenting by deformation of the size of each image with small changed in ground-truth rectangles corners coordinates do not increased robustness of detection, or quality. Some results, that not represented here, show that increase of contrast and sharpness of testing set results in higher quality of detection. Influence of filtering will be tested more thoroughly in our future work.

Attempt to use simpler architecture with different layers' size resulted in lower detection quality, so in the future works we will test last of represented architects, that should result in best quality of detection.

References

1. Stepanov, D., Bakhshiev, A., Gromoshinskii, D., Kirpan, N., Gundelakh, F.: Determination of the relative position of space vehicles by detection and tracking of natural visual features with the existing TV-cameras. Analysis of Images, Social networks and Texts, Four International Conference, AIST 2015, Yekaterinburg, Russia, 9–11 April 2015, Revised Selected papers. Communications in Computer and Information Science, vol. 542, pp. 431–442 (2015)
2. Bakhshiev, A.V., Korban, P.A., Kirpan, N.A.: Software package for determining the spatial orientation of objects by TV picture in the problem space docking. In: Robotics and Technical Cybernetics, Saint-Petersburg, Russia, RTC, vol. 1, pp. 71–75 (2013)
3. InterSpace. The channel is hosting a record of all space launches in the world. https://www.youtube.com/channel/UC9Fu5Cry8552v6z8WimbXvQ. Accessed 19 Apr 2017
4. Ren, S., He, K., Girshick, R., Sun, J.: Faster R-CNN: towards real-time object detection with region proposal networks. In: Neural Information Processing Systems (NIPS) (2015)
5. Girshick, R.: Faster R-CNN (Python implementation). https://github.com/rbgirshick/py-faster-rcnn
6. Chatfield, K., Simonyan, K., Vedaldi, A., Zisserman, A.: Return of the Devil in the Details: Delving Deep into Convolutional Nets. British Machine Vision Conference (2014)
7. Model Zoo – BVLC. affe Wiki – GitHub. https://github.com/BVLC/caffe/wiki/Model-Zoo
8. Everingham, M., Van Gool, L., Williams, C.K.I., Winn, J., Zisserman, A.: The Pascal Visual Object Classes (VOC) Challenge. http://host.robots.ox.ac.uk/pascal/VOC/pubs/everingham10.pdf

Detection of Neurons on Images of the Histological Slices Using Convolutional Neural Network

Ivan Fomin[1]([⊠]), Viktor Mikhailov[1], Aleksandr Bakhshiev[1],
Natalia Merkulyeva[2,3], Aleksandr Veshchitskii[3],
and Pavel Musienko[2,4,5]

[1] The Russian State Scientific Center for Robotics and Technical Cybernetics,
Tikhoretsky Prospect 21, 194064 Saint Petersburg, Russia
{i.fomin,v.mikhaylov,alexab}@rtc.ru
[2] Institute of Translation Biomedicine, Saint-Petersburg State University,
194064 Saint Petersburg, Russia
mer-natalia@yandex.ru, pol-spb@mail.ru
[3] Pavlov Institute of Physiology RAS, Saint Petersburg, Russia
veschickiyalex@mail.ru
[4] Saint-Petersburg State University of Aerospace Instrumentation,
67, B. Morskaia Street, 190000 Saint Petersburg, Russia
[5] Children's Surgery and Orthopedic Clinic,
Department of Nonpulmonary Tuberculosis,
Institute of Phthysiopulmonology, Saint Petersburg, Russia

Abstract. An automatic analysis of images of the histological slices is one of main steps in process of description of structure of neural network in norm and pathology. Understanding of structure and functions of that networks may help to improve neuro-rehabilitation technologies and to translate experimental data to the clinical practice. Main problem of the automatic analysis is complexity of research object and high variance of its parameters, such as thickness and transparency of slice, intensity and type of histological marker, etc. Variance of parameters make every step of neuron detection very hard and complex task. We represent algorithm of neuron detection on images of spinal cord slices using deep neural network. Networks with different parameters are compared to previous algorithm that based on pixels' filtration by color.

Keywords: Neuron detection · Deep learning · Object detection · Image segmentation · Neural networks · Spinal cord slices · NeuN

1 Introduction

Development of new tools for experimental data analysis is important task to increase the effectiveness of studies devoted to the neuronal control of locomotor and postural functions. Investigations in this area are significant for the invention of new rehabilitation methods of peoples with neuromotor disorders [1–3]. Essential step in these studies is creation of maps of neuronal distribution within the spinal cord.

© Springer International Publishing AG 2018
B. Kryzhanovsky et al. (eds.), *Advances in Neural Computation, Machine Learning, and Cognitive Research*, Studies in Computational Intelligence 736,
DOI 10.1007/978-3-319-66604-4_13

Cell detection on images of the histological slices is an intensively developing area of interest in bioinformatics. But widely used program solutions [4, 5] are not able to detect neurons on images containing some defects. There are some articles describing algorithms of neuron detection on both normal and defective images [6–9], but these algorithms were developed for 3D images created using high-resolution electronic microscope. That algorithms cannot be applied to 2D images of histological slices that contains many different physical defects, so other approach is required.

Last decade (since 2006 year) the popularity of the deep convolutional neural networks grows exponentially. These algorithms have been applied for the data and image analysis in many different tasks like object detection [10, 11], object classification [12], and image generation for some scientific applications or for entertainment [13]. Convolutional networks have been also applied for semantic segmentation tasks, and have shown good results on PASCAL VOC dataset [14]. The main advantage of the neural networks over the classical image processing methods is requirement only the initial images and the ground truth markup, that relied upon the current task. Classical (image segmentation) methods needs large set of rules to segment different areas of the image, set must be hand-made for each area. Main disadvantage of convolutional neural network is need in the large training dataset, and in the learning dataset covered all or almost all possible variations in the training dataset (it is especially difficult to achieve when histological slices are processed).

An automatic neuron detection in histological slice images is very hard because of different problems. (1) Some neurons are seem to be more pale than others because of their location out of focus; this is a reason of artifacts in optical density assessment. (2) Not only soma of neurons, but also fragments of their processes (axon, dendrites) are detected by some neuromorphological methods; potentially this may cause false results of soma size detection or cells counting. (3) Images of neurons may contain particular physical defects of the tissue: fractures, compressions, etc., that complicates an automatic detection of the objects of interest. (4) In many histological methods one image can contain neurons of very different sizes, that makes harder the tuning of parameters for automatic detection of neurons on the slice. (5) Different photo capturing conditions such as brightness of light source, opacity of glasses around slice and slice itself can cause shift in the intensity of RGB channels of the image and can results in additional complication in the development of algorithm for automatic neuron detection. Examples of fragments of histological slice images represented on Fig. 1.

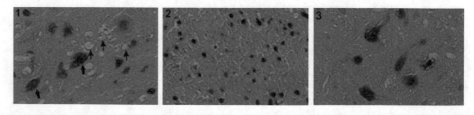

Fig. 1. Fragments of cat spinal cord slices. Immunohistochemical detection of the NeuN protein. All three fragments (1, 2 and 3) are extracted from different regions of the same slice. Thick arrows point to the bodies of neurons, thin arrows indicate slice defects

The problem of automatic detection of the neurons on histological slices images is non-trivial and requires complex algorithm allowing to solve all difficulties described above. The aim of the present work is a development and testing of such algorithm based upon the neural-network approach.

2   Methods

Figure 1 contains examples of fragments of one of the frontal slices of cat spinal cord after immunohistochemical detection of special neuronal marker – NeuN protein. To solve problem of automatic neuron detection on slices we developed deep neuronal network (see Fig. 2) that is Fully Convolutional Network (FCN) [10, 11].

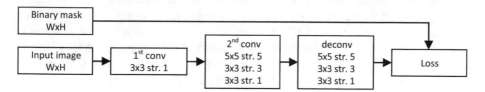

Fig. 2. Overall scheme of the network.

First convolutional layer contains 100 filters 3×3, parameters of layer are applied so that output of layer has size equal to input image. This layer developed for detection of the simple structural elements, like borders and corners. Second layer contains 50 filters of size 3×3 or 5×5, and is different for each test. Small size of filter in second layer chosen in assumption that decision if pixel rely to neuron or not can be done using color of pixel and its small surrounding; experiments approved that assumption is right.

In first two tests second layer work as grid of small fully-connected networks with each cell 3×3 and 5×5 respectively, each small "network" works in its own cell and returns respective output. In third test filter applied to each area of the image as sliding window, experiments showed that this approach with filter size 3×3 returns better result both visually and numerically.

Third layer is deconvolutional layer that uses set of filters learned by second convolutional layer to receive image of source size. Last layer (loss) compare output of deconvolutional layer to target image where areas where are neurons marked with white color, and other areas marked with black. Target of learning is to minimize loss between output of deconvolutional layer and target image.

3   Results

We pre-processed the learning/testing and verification datasets. Each dataset contains fragments of images of the slices (see Fig. 3, left) and corresponding binary masks where white areas are neurons and black areas are regions surrounded them (surrounding regions) (see Fig. 3, right).

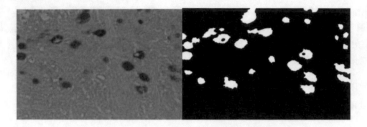

Fig. 3. Example of learning/testing markup

To verify the results of network and to compare them to previously developed algorithm (based on LAB image transform and adaptive filtering by histograms of components of the transform [15]) we decided to find rectangles around white areas in markup, and every resulting binary image. To verify the detection, we used two criteria: Intersection over Union (IoU) for rectangle in ground truth image and detection result more than 50% or distance between centers of rectangles less than 13 pixels. Second criteria is required because neural network can detect neuron properly, but size of white area will be less that enough to achieve IoU 50% and more.

Table 1. Comparison of precision and recall for different approaches

	LAB filter		5 × 5, step 5		3 × 3, step 3		3 × 3, step 1	
	Prec	Rec	Prec	Rec	Prec	Rec	Prec	Rec
39	0.8	0.909	0.941	0.762	1.000	0.571	1.000	0.571
42	0.944	1.0	0.536	0.882	0.750	0.882	0.714	0.882
44	0.765	1.0	0.733	0.846	0.889	0.615	0.889	0.615
46	0.889	1.0	0.941	0.941	0.923	0.750	0.857	0.750
49	1.0	1.0	0.632	0.727	0.913	0.636	0.864	0.576
50	0.867	0.975	0.909	0.750	1.000	0.675	1.000	0.675
51	0.762	0.762	0.739	0.810	0.800	0.762	0.762	0.762
53	0.765	0.722	0.706	0.667	0.818	0.529	0.818	0.500
54	0.957	0.957	0.840	0.913	1.000	0.762	1.000	0.773
55	0.719	0.885	0.882	0.577	0.667	0.231	0.750	0.231
56	0.952	1.000	0.611	0.550	0.667	0.300	0.571	0.200
57	0.933	0.933	0.743	0.867	0.730	0.900	0.730	0.871
58	0.92	1.000	0.630	0.739	0.905	0.826	0.900	0.783
60	0.882	0.833	0.818	0.500	0.571	0.222	0.625	0.278
62	0.909	0.976	0.884	0.950	1.000	0.927	0.973	0.900
63	0.871	0.964	0.806	0.926	0.923	0.889	0.958	0.852
64	0.885	0.920	0.909	0.800	0.895	0.630	0.684	0.520
65	0.913	0.840	0.947	0.720	1.000	0.538	0.929	0.500
66	0.828	0.923	0.905	0.731	0.882	0.556	0.889	0.615

Using this two approaches, we could calculate count of true positives (successfully detected, TP), false positives (false detections, FP), and false negatives (missed neurons, FN) for previous method and each configuration of neural network. Using this data, we calculated precision (Prec) = TP/(TP + FP) and recall (Rec) = TP/(TP + FN) for each methods and collected them to Table 1. In the left column represented number of image in the set, each pair of columns contains precision and recall for each method, that subscribed in top row.

Best precision has been received with network where second layer contains convolution with 3×3 filter with step 1. Best recall (detected more neurons) we received using previously developed method with LAB filter. Main reason recall of method based on neural network is not enough is small size of learning set (not all possible neuron views are covered).

4 Conclusions

An approach for detection neurons on photo images of histological slices of cat spinal cord using deep learning fully-convolutional neural network is represented. Results shown precision of detection up to 88% for one fragment that outperforms precision of previously developed approach based on LAB image transform.

Disadvantage of the presented approach is lower completeness of detection. To improve it, in future we will increase the size of learning data set, and will combine the results of networks computation with different architects to not lost neurons that hard for one separate architect. Also we plan to apply to this task more networks especially developed for object segmentation.

Acknowledgements. This work performed with financial support of Russian Science Foundation by Grant № 14-15-00788.

References

1. Musienko, P., Heutschi, J., Friedli, L., den Brand, R.V., Courtine, G.: Multi-system neurorehabilitative strategies to restore motor functions following severe spinal cord injury. Exp. Neurol. **235**(1), 100–109 (2012)
2. Musienko, P., van den Brand, R., Maerzendorfer, O., Larmagnac, A., Courtine, G.: Combinatory electrical and pharmacological neuroprosthetic interfaces to regain motor function after spinal cord injury. IEEE Trans. Biomed. Eng. **56**(11), 2707–2711 (2009)
3. Bakhshiev, A.V., Smirnova, E.Y., Musienko, P.E.: Methodological bases of exobalancer design for rehabilitation of people with limited mobility and impaired balance maintenance. Izv. SFedU. Eng. Sci. **10**(171), 2011–2013 (2015)
4. Kolodziejczyk, A., Habrat (Ladniak), M., Piorkowski, A.: Constructing software for analysis of neuron, glial and endothelial cell numbers and density in histological Nissl-stained rodent brain tissue. J. Med. Inform. Technol. **23**, 77–86 (2014)
5. Al-Kofahi, Y., Lassoued, W., Lee, W., Roysam, B.: Improved automatic detection and segmentation of cell nuclei in histopathology images. IEEE Trans. Biomed. Eng. **57**(4), 841–852 (2010)

6. Dong, B., Shao, L., Da Costa, M., Bandmann, O., Frangi, A.F.: Deep learning for automatic cell detection in wide-field microscopy zebrafish images. In: 2015 IEEE 12th International Symposium on Biomedical Imaging (ISBI), pp. 772–776 (2015). ISSN 1945-7928
7. Karakaya, M., Kerekes, R.A., Gleason, S.S., Martins, R.A., Dyer, M.A.: Automatic detection, segmentation and characterization of retinal horizontal neurons in large-scale 3D confocal imagery. SPIE Med. Imaging (2011). doi:10.1117/12.878029
8. Lin, G., Chawla, M.K.: A multi-model approach to simultaneous segmentation and classification of heterogeneous populations of cell nuclei in 3D confocal microscope images. Cytometry Part A **71**(9), 724–736 (2007)
9. Oberlaender, M., Dercksen, V.J., Egger, R., Gensel, M., Sakmann, B., Hege, H.C.: Automated three-dimensional detection and counting of neuron somata. J. Neurosci. Methods **180**(1), 147–160 (2009). doi:10.1016/j.jneumeth.2009.03.008. Epub 21 Mar 2009
10. Ren, S., He, K., Girshick, R., Sun, J.: Faster R-CNN: towards real-time object detection with region proposal networks. In: Advances in Neural Information Processing Systems (NIPS) (2015)
11. Redmon, J., Farhadi, A.: Yolo9000: better, faster, stronger. arXiv preprint, arXiv:1612. 08242 (2016)
12. Chatfield, K., Simonyan, K., Vedaldi, A., Zisserman, A.: Return of the devil in the details: delving deep into convolutional nets. In: British Machine Vision Conference (2014). arXiv: 1405.3531
13. Isola, P., Zhu, J.-Y., Zhou, T., Efros, A.A.: Image-to-image translation with conditional adversarial networks. In: Computer Vision and Pattern Recognition (cs.CV), submitted 21 Nov 2016. arXiv:1611.07004[cs.CV]
14. Shelhamer, E., Long, J., Darrell, T.: Fully convolutional models for semantic segmentation. In: PAMI (2016). arXiv:1605.06211
15. Mikhaylov, V.V., Bakhshiev, A.V.: The system for histopathology images analysis of spinal cord slices. Procedia Comput. Sci. **103**, 239–243 (2017)

Constructing a Neural-Net Model of Network Traffic Using the Topologic Analysis of Its Time Series Complexity

N. Gabdrakhmanova$^{(\boxtimes)}$

People's Friendship University of Russia, Moscow, Russia
gabd-nelli@yandex.ru

Abstract. The dynamics of data traffic intensity is examined using traffic measurements at the interface switch input. The wish to prevent failures of trunk line equipment and take the full advantage of network resources makes it necessary to be able to predict the network usage. The research tackles the problem of building a predicting neural-net model of the time sequence of network traffic.

Topological data analysis methods are used for data preprocessing. Nonlinear dynamics algorithms are used to choose the neural net architecture. Topological data analysis methods allow the computation of time sequence invariants. The probability function for random field maxima cannot be described analytically. However, computational topology algorithms make it possible to approximate this function using the expected value of Euler's characteristic defined over a set of peaks. The expected values of Euler's characteristic are found by constructing persistence diagrams and computing barcode lengths. A solution of the problem with the help of R-based libraries is given. The computation of Euler's characteristics allows us to divide the whole data set into several uniform subsets. Predicting neural-net models are built for each of such subsets. Whitney and Takens theorems are used for determining the architecture of the sought-for neural net model. According to these theorems, the associative properties of a mathematical model depend on how accurate the dimensionality of the dynamic system is defined. The sub-problem is solved using nonlinear dynamics algorithms and calculating the correlation integral. The goal of the research is to provide ways to secure the effective transmission of data packets.

Keywords: Computational topology · Persistence · Stability · Neural network

1 Introduction

The topicality of the study is determined by the following reasons. The continuing development of telecommunication and Internet services sets new requirements for the bandwidth of telecommunication channels. The presence of a great deal of various services in a single physical transmission medium at pick hours can bring about the overloading of switching and routing devices in trunk lines and, as result, a reduction of many services. The wish to prevent failures of trunk line equipment and take the full advantage of network resources makes the problem of effective use of the

© Springer International Publishing AG 2018
B. Kryzhanovsky et al. (eds.), *Advances in Neural Computation, Machine Learning, and Cognitive Research*, Studies in Computational Intelligence 736,
DOI 10.1007/978-3-319-66604-4_14

telecommunications channel bandwidth very important (the direct widening of the bandwidth inevitably leads to an increase of service costs). It is necessary to have effective traffic control methods that could use statistical data to predict the traffic intensity. A lot of modern publications deal with mathematical models of different types of network traffic [1–3]. The complexity and relevance of this problem urge further research in the field.

2 The Topological Data Analysis

The topological data analysis is a new theoretical trend in the field of data analysis. The approach allows the determination of topological data structures. Recent advancements in the field of computational topology make it possible to find topological invariants in data collections [2, 4, 5].

The point of the analysis is that stable properties are to be immune to noise, distortions, errors, lack of data. The practice of using the analysis in different fields shows that the supposition is true and stable topological properties can provide a lot of information about data collections. Persistence diagrams are one of basic tools of computational topology. They make it possible to get useful information about excursion sets of a function. Below are the basic definitions (according to [4]).

Let X be a topological space being triangulated, f be a continuous tame function defined over space X. Let us introduce the notation $U_a = f^{-1}(-\infty, a]$ for $a \in R$. When moving upwards, components U_a can merge or produce new components. It is possible to trace how the sub-level topology changes with a by examining homologies of these sets with, say, persistence homologies. Parameter $a \in R$ is called the homological critical value if for certain k the homomorphism induced by nesting $f_* : H_k(U_{a-\varepsilon}) \to H_k(U_{a+\varepsilon})$ is not an isomorphism for any sufficiently small $\varepsilon > 0$ (homology groups are considered with coefficients in Z_2). Continuous function f is called tame function if it has a finite number of homological critical values. When $b \leq a$, then $U_b \subseteq U_a$. Let us denote a set of connectivity components as $C(a) = C(U_a)$. It is possible to define a functional – Euler characteristic – over a set of sub-levels of U_a. Let $X \subset R^2$. Then, in the terms of algebraic topology, Euler's number is $\chi(U_a) = \beta_0 - \beta_1$, where β_0, β_1 are the ranks of the first two homology groups. Functional $\chi(U_a)$ measures the field topological complexity on the sub-level set. Note that for function f it is possible to deal with a set of super-levels $U_a = f^{-1}[a, \infty)$ instead of sub-levels.

Let us define the persistence diagram according to [5]. Let $f\colon X \to R$ be a tame function. Let $a_1 < a_2 < \ldots < a_n$ be critical homological values. Let us consider interjacent values $b_0, b_1, \ldots, b_n : b_{i-1} < a_i < b_i$. Let us supplement the chosen points in the following way: $b_{-1} = a_0 = -\infty$; $b_{n+1} = a_{n+1} = +\infty$. Let us define the multiplicity of point $(a_i < a_j)$ for each couple of indices $0 \leq i < j < n + 1$ by setting $\mu_i^j = \beta_{b_{i-1}}^{b_j} - \beta_{b_i}^{b_j} + \beta_{b_i}^{b_{j-1}} - \beta_{b_{i-1}}^{b_{j-1}}$, where $\beta_x^y = \dim(\mathrm{Im}(f_x^y)), f_x^y : H_k(U_x) \to H_k(U_y)$. Persistence diagram $D(f) \subset R_2$ of function f stands for a set of points (a_i, a_j) $(i, j = 0, 1, \ldots, n + 1)$ adjusted for multiplicity μ_i^j in combination with a set of diagonal points $\Delta = \{(x, x) | x \in R\}$ adjusted for infinite multiplicity.

The immunity of a persistence diagram to perturbations of function f is its remarkable feature. Persistence diagrams can be used to calculate the lengths of the barcodes of connectivity components. Here the term barcode stands for the component lifetime. Let us denote the summarized lengths of barcodes of two homology groups H_0 and H_1 as L_0 and L_1 correspondingly. Then the mean of the Euler characteristic can be determined [2] as

$$\chi = L_0 - L_1. \tag{1}$$

3 Setting the Problem

A second-level interfacial switch of a backbone line provider is taken as a test object in the paper. The traffic coming to each port of the switch is integrated traffic from user groups belonging to a particular region. The explanatory drawing is given in Fig. 1. The Cacti software (SNMP interface protocol) was used to gather statistic data. The information about the degree of network usage is more useful in practice. The knowledge of the number of packets in unit time can be misleading. For this reason the aggregate quantity $x(t)$ – traffic intensity (in bits) at moment t – is taken as an observable variable. The extension of data is 10080 points or 7 days. The plot of traffic intensity measured at port GE 0 is shown in Fig. 2. Each point in this plot represents a number of bits going through the trunk in one minute's time.

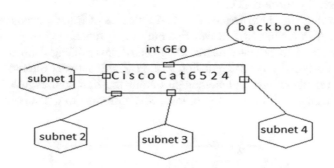

Fig. 1. The measurement arrangement.

So the goal is to construct a mathematical model for the m-step prediction of traffic intensity using observations $\{x(t), t = 1, 2, ..., N\}$, where N is the number of points. The estimates of Euler's characteristics are used here as indication of network usage. The following algorithm is proposed. The whole data collection is to be divided in clusters with different Euler's characteristics. A neural-net prediction model is to be built for each cluster using nonlinear dynamics methods. Below is the result of the experimentation.

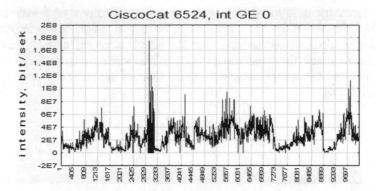

Fig. 2. The traffic intensity plot at port GE 0.

4 Topological Invariants Calculated for a Traffic Intensity Sequence

Packet TDA from a public repository of R packets was used as a library for finding stable homologies. The packet has a broad toolkit for topological data analysis by topological methods.

Before finding topological characteristics, the whole data collection was divided in some portions. Each portion held data acquired in two hours' time. For each portion persistence diagrams, barcodes were determined and Euler's characteristic estimates were calculated by formula (1).

The following algorithm was used to find estimates of Euler's characteristic in the TDA packet. A triangulation grid was first built using function Grid(). Then function gridDiag was used to produce matrix Diag. Function gridDiag evaluates the actual value of the function by the triangulation grid, generates simplex filtration using these values, and calculates constant homologies from the filtration. Figure 3 shows the persistence diagrams for one portion of data. The birth time of a component is plotted

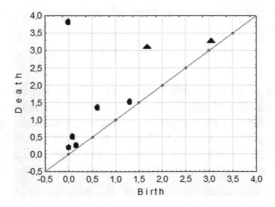

Fig. 3. The plot on the right shows the persistence diagram of the superlevel sets of the KDE.

as abscissas; the death time is plotted as ordinates. The dots correspond to zero-dimensional simplexes, the triangles mark single-dimensional simplexes. Figure 4 presents the barcode chart of zero-dimensional simplexes. Table 1 gives the estimates computed for different (n = 15) portions of the object. The following notation is used in the table: n is the number of a portion (interval), L0 and L_1 are the summarized barcode lengths of zero- and single-dimensional simplexes, χ is the estimate of Euler's characteristic (1). The plot in Fig. 5 shows Euler's characteristic as function of n. The horizontal axis represents the number of an interval and Euler's characteristic is measured on the vertical axis.

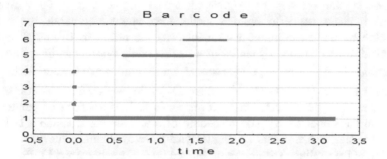

Fig. 4. Barcode

Table 1. The estimates of Euler's characteristic

n	1	2	3	4	5	6	7	8	9	10	11	12
L_0	3.7	4.1	3.6	3.7	2.2	1.7	2.0	2.0	1.8	2.4	3.4	3.6
L_1	2.3	1.6	1.7	1.8	2.5	1.7	2.0	2.0	1.5	3.3	1.8	2.3
χ	1.5	2.6	1.9	1.8	−0.3	−0.1	0.03	0.04	0.3	−0.8	1.5	1.4

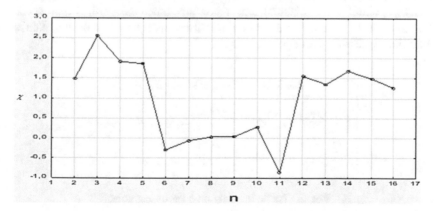

Fig. 5. Euler's characteristic as function of n.

The results prove that Euler's numbers are a stable characteristic of traffic intensity. At the next stage the portions with the same $[\chi]$ (where $[.]$ is the integer of a number) are united in a single cluster. A neural-net prediction model is built for each cluster.

5 Building the Neural-Net Model of the Data

Methods of nonlinear dynamics are used to construct a neural-net model for a selected cluster. The subproblem is set as follows. Let $\{x(t)\}_{t=1}^{N}$ be measurements of a particular observable scalar component of a d_1-dimensional dynamic system $\bar{y}$. On the whole, the dimensionality and behavior of the dynamic system are not known. For a given time sequence it is necessary to build a model that would incorporate the dynamics responsible for the generation of observations $x(t)$. According to Takens' theorem, the geometrical structure of the dynamics of a multivariable system can be restored using observations $\{x(t)\}_{t=1}^{N}$ in a D-dimensional space built around new vector $\bar{z}(t) = \{x(t), x(t-1), \ldots, x(t-(D-1))\}^{T}$ (where $D \geq 2d_1 + 1$). The evolution of points $\bar{z}(t) \rightarrow \bar{z}(t+1)$ in the restored space corresponds to the evolution of points $\bar{y}(t) \rightarrow \bar{y}(t+1)$ in the initial space. The procedure of searching for a suitable D is called nesting. The least value of D at which the dynamic restoration is achieved is called the dimension of the nesting. The algorithm offered by P. Grassberger and I. Proccaccia in 1983 makes it possible to evaluate D using a time sequence.

After D is estimated, the problem at hand can be formulated in the following way. There is time series $\{x(t)\}_{t=1}^{N}$ and restoration parameters ($D = 11$ in our case) are set. For N_1 vectors $\bar{z}(t) = \{x(t), x(t-1), \ldots, x(t-(D-1))\}^{T}$ the values of the sought-for function $F(t) = F(\bar{z}(t))$ are known (because the terms of the time series following $\bar{z}(t)$ are known). It is necessary to find the value of the sought-for function at new point $\bar{z}(t)$, $\hat{x} = F(\bar{z})$.

Neural nets of the multiple-layer perceptron type [6] are used to tackle the problem. Only the key results are given below. Figure 6 shows the graph of traffic intensity on a set of test points. The horizontal axis represents time, the vertical axis shows the normalized traffic intensity; the solid line corresponds to experimental data x, the dashed line represents theoretical results $\hat{x}$.

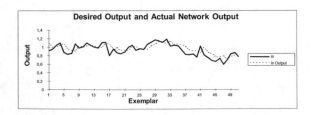

Fig. 6. Traffic intensity on a set of test points

6   Conclusions

The goal of the paper was to test the hypothesis that the use of the topological data analysis would make it possible to build traffic intensity prediction models due to finding additional characteristics that cannot be discovered by conventional data analysis. The data of network traffic intensity in a week's time were examined. The computations showed that the traffic intensity dynamics can be described by Betti numbers and Euler's characteristics. The algorithm using Euler's characteristics was used in the paper to build a model makes it possible to increase the prediction accuracy by an order of magnitude (as compared with methods not using Betti numbers). The paper gives the results of first steps towards the application of topological data analysis for predicting the network traffic intensity. The results proved the prospectiveness of further research in the field.

References

1. Heyman, D.P., Tabatabai, A., Lakshman, T.V.: Statistical analysis and simulation study of video teleconference traffic in ATM networks. IEEE Trans. Circuits Syst. Video Technol. **2**, 49–59 (1992)
2. Zhani, M.F., Elbiaze, H.: Analysis and prediction of real network traffic. J Netw **4**(9), 855–865 (2009)
3. Potapov, A.B.: Time-series analysis: when dynamical algorithms can be used. In: Proceedings of 5th International Specialist Workshop Nonlinear Dynamics of Electronic Systems, Moscow, 26–27 June 1997, pp. 388–393 (1997)
4. Edelsbunner, H., Letsscher, D., Zomorodian, A.: Topological persistence and simplification. Discret. Comput. Geom. **28**, 511–533 (2002)
5. Carlsson, G., Zomorodian, A.: Computing persistent homology. In: Proceedings of 20th Annual Symposium on Computational Geometry, pp. 347–356 (2004)
6. Haykin, S.: Neural Networks: A Comprehensive Foundation, 2nd edn. (2006)

Texture Recognition from Positions of the Theory of Active Perceptions

Vasiliy Gai$^{(\boxtimes)}$, Pavel Rodionov, Maxim Derbasov, Dmitriy Lyakhmanov, and Alla Koshurina

Nizhny Novgorod State Technical University N. A. R. E. Alekseev,
K. Minin Street, 24, Nizhny Novgorod 603950, Russian Federation
vailiy.gai@gmail.com

Abstract. Recognition of textures is one of the topical tasks of computer vision. The key step in solving this problem is the formation of feature description of the texture image. A new approach to the formation of texture features based on the theory of active perception is proposed. The results of a computational experiment based on the Brodatz-32 database are presented, and the accuracy of the classification is demonstrated. The application of the proposed feature systems for recognition of snow and land textures in the solution of the problem of auto piloting in complex natural and climatic conditions is considered.

Keywords: Texture recognition · Theory of active perception · Auto piloting

1 Introduction

The task of recognizing textures is one of the fundamental problems in the field of computer vision and image processing. Texture recognition is used for automatic and automated analysis of medical images, object recognition, environmental modeling and image search in databases. The methods for recognizing textures also find their application in solving the problem of auto piloting in complex natural and climatic conditions.

The structure of the system for recognizing textured images can be represented as a set of three stages: image preprocessing, formation of image feature system and making decision.

The preliminary processing of the image usually consists in applying to the image of the filter suppressing noise. Often the implementation of this stage is not fulfilled. In this case, the responsibility for the noise invariance is shifted to the method of forming feature description.

When forming the feature description of a textured image, a wavelet transform is used, a method of analyzing independent components, Markov random fields and so on. In [1], it is proposed to use the Gabor filter bank to calculate the feature description. However, the filters included in the Gabor filter bank

© Springer International Publishing AG 2018
B. Kryzhanovsky et al. (eds.), *Advances in Neural Computation, Machine Learning, and Cognitive Research*, Studies in Computational Intelligence 736,
DOI 10.1007/978-3-319-66604-4_15

are not orthogonal. The method of forming an image description based on local binary patterns is presented in [2]. It is based on the calculation of the sign of the difference between the brightness of neighboring samples. This method is not resistant to noise due to the use of threshold operation. The modification of the method of local binary patterns is known – the method of local ternary patterns [3]. The description formed on the basis of local ternary patterns is noise-resistant, but not resistant to brightness level changes, since the algorithm for generating the feature description uses fixed predetermined thresholds.

When solving the problem of classification of textured images on the basis of a well-known feature description, the method of K-nearest neighbors and artificial neural networks are often used.

This paper is devoted to solving the problem of recognizing textured images from the perspective of the theory of active perception (TAP).

2   Using TAP in Image Recognition

The basic transformation of TAP is a U-transformation, which is realized in two stages [4]. In the first stage, the Q-transformation is applied to the image, after which we obtain a matrix of visual masses m with a size of 4×4 elements. In the second stage, the set of filters $F = \{F_i\}$ is applied to the result of the Q transformation.

The filter element can take the values "+1" (dark areas) and "−1" (light areas). Structurally, these filters are similar to the Walsh filters of the Harmuth system. The specificity of using these filters is that they are applied after the implementation of the Q-transformation.

In the TAP, to each filter F_i the binary operator V_i is put in correspondence. In this case the operator V_i or $\overline{V_i}$ corresponds to the component $\mu_i \neq 0$ of the vector μ_i depending on the sign of the component.

Defining the operations of set-theoretic multiplication and addition on the set $\{V_i\}$, an operation analogous to negation, two elements: $1 - V_0 \ 0 - \overline{V_0}$ we obtain the algebra of signal description in Boolean functions: $AV =< \{V_i\} : +, \times >$. For any $V_i, V_j, V_k \in V$, the laws of commutativity, associativity, idempotency, and distributivity are fulfilled.

A group algebra is formed on the set of operators:

1. the family of algebraic structures P_n of cardinality 35, called complete groups, are formed on triples of operators (V_i, V_j, V_k);
2. the family of algebraic structures P_s of power 105 (for 16 filters), called closed groups, is formed on the four operators (V_i, V_j, V_p, V_m);

Comparing TAP with the known approaches to the formation of the feature description of the signal, we can note the following:

1. in comparison with the wavelet transform and the Fourier transform, TAP makes it possible to calculate, with respect to spectral coefficients, signs of a higher level (due to the use of group algebra);

2. in comparison with the models of deep learning in TAP, the feature description is calculated without using training, but by predefined templates;
3. only the addition and subtraction operations are used in calculating the U-transformation.

3   Formation of Feature Description of a Textured Image

The algorithm for forming the feature description of an image consists in combining descriptions of individual areas of the original image, obtained on the basis of complete or closed groups, into histograms of complete and closed groups. The algorithm for forming the feature description can be written as follows:

$$\forall i = 1 : s_h : (N - h)$$
$$\quad \forall j = 1 : s_w : (M - w)$$
$$\quad\quad I_s = I[i : (i + s_h - 1); j : (j + s_w - 1)];$$
$$\quad\quad I_G = H[I_G, \Gamma];$$
$$\quad\quad I_D[I_G] = I_D[I_G] + 1.$$

The following notations are used in the algorithm record: I_D – the image description obtained during the operation of the algorithm, I_s is the region of the image over which the I_G description is formed, the size of the region I_s – $h \times w$ samples, Γ is the type of the description being formed: P_{nm} – complete groups on the operation of multiplication, P_{na} – complete groups on the addition operation, P_s – closed groups. The I_G description can be obtained in the form of complete or closed groups. The value of the shift step of the area I_s in the

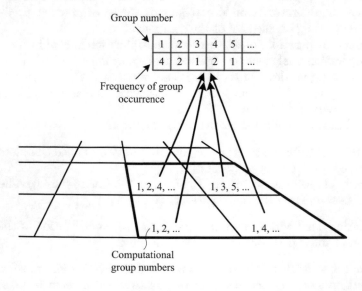

Fig. 1. The scheme of calculation of the histogram of groups

Table 1. Results of the use of histograms of complete and closed groups

Number of images in the sample	Feature descriptions used	Image area size I_s	Accuracy of classification (in %)	Time of feature description calculation (in ms)
$32 \times 16 = 512$	P_s	64×64	97	4
$32 \times 36 = 1024$	$P_s + P_{na}$	32×32	97	10
$32 \times 64 = 2048$	$P_s + P_{nm} + P_{na}$	32×32	94	5

image I is s_h of horizontal and s_w vertical counts of samples. The realization of the operator H for the calculation of complete and closed groups by the image is given in [4].

Figure 1 shows the scheme for calculating the group histogram.

4   Computational Experiment

As the initial data for the implementation of the computational experiment, the Brodatz-32 texture image database is used. When solving the classification problem, the support vector method with a linear kernel is used.

The results of estimating the accuracy of classification using various feature descriptions are given in Table 1. To form a sample, each texture image is divided into 16, 32 and 64 equal parts. The values of the parameters s_h and s_w are 1/4 of the size of the area.

The results of testing the proposed approach to the formation of the feature description to various distortions of the classified image are given in Tables 2 and 3. As the feature description, the description obtained after merging the histograms of complete (on addition and multiplication) and closed groups was used. This feature description provides the best results when classifying textured images without distortion.

The size of the classified image is 256×256 samples, the size of the sample is 512 images, the value of the window shift in the s_w (s_h) image is 32 samples. During the testing, as a test sample the distorted images were used, and as a training sample – the entire database of images of textures.

Table 2 shows the results of the evaluation of classification accuracy when the classified image is rotated by a given angle. Table 3 shows the results of experiments to study the invariance of the proposed feature description to the distortion of the classified image by normal noise.

Table 4 shows the results of the classification of the Brodatz-32 database on the basis of the known methods. Table 5 shows the evaluation of the noise invariance of the known systems of features (in %).

Table 2. Evaluation of the invariance to rotation

Angle of rotation (in degrees)	1	2	3	4	5	6	7	8	9	10	15	20
Accuracy of classification (in %)	94	93	93	89	93	91	90	89	87	87	70	58

Table 3. Evaluation of the noise invariance

Signal to noise ratio (in decibels)	20	15	10	5	0
Accuracy of classification (in %)	92	68	59	31	15

Table 4. Accuracy of classification based on the known methods

Reference	Feature description of the texture image	Algorithm of classification	Accuracy (in %)
[2]	Gabor filters	KNN	96
[2]	Local binary patterns	KNN	98
[3]	Convolutional neural network	Convolutional neural network	91
[5]	Local ternary patterns	SVM	95
[5]	Local high order statistics	SVM	99

Table 5. Invariance to distortion by noise of the known feature descriptions

Invariance to distortion by noise/Feature descriptions	20	10	5
SIFT	87	55	32
LBP	85	53	20

Comparing the proposed approach with the known ones, we can note the following:

1. the filters used in constructing the feature description of a textured image, in contrast to Gabor filters, are orthogonal;
2. when forming a feature description, in comparison with Local Binary Patterns and Local Ternary Patterns, there is no need to set predefined thresholds;
3. the invariance of the proposed feature descriptions to the distortion of the classified image by normal noise is comparable to the invariance to normal noise of LBP and SIFT features [2];
4. in [2] it is indicated that the average time for calculating LBP features by the image is 0.6 ms, based on Gabor filters – 197 ms (Intel Core i5-2400, C++); The time of calculation of the proposed systems of features depends on the size of the processed image, the size of the I_s region, and the magnitude of the

shift of the I_s region over the image, and ranges from 1 to 19 ms (Intel Core i7-4790K, C++); Thus, by the speed of computation the proposed feature descriptions are as good as the known ones.

5 Conclusion

The article considers an approach to the formation of the feature description of a textured image based on the theory of active perception. The practical application of this feature description is the segmentation of the image obtained from the video camera to solve the tasks of autopilot in the absence of a road network. Testing of the proposed approach to the formation of the feature description is performed using the method of reference vectors based on the Brodatz-32 textured image database. We obtained the result showing the effectiveness of the proposed systems of characteristics, their invariance to various distortions of the classified images.

Acknowledgements. The work was carried out at the NNSTU named after R. E. Alekseev, with the financial support of the Ministry of Education and Science of the Russian Federation under the agreement 14.577.21.0222 of 03.10.2016. Identification number of the project: RFMEFI57716X0222. Theme: "Creation of an experimental sample of an amphibious autonomous transport and technological complex with an intelligent control and navigation system for year-round exploration and drilling operations on the Arctic shelf."

References

1. Rosniza, R., Nursuriati, J.: Texture feature extraction using 2-D Gabor Filters. In: IEEE Symposium on Computer Applications and Industrial Electronics, ISCAIE 2012, pp. 173–178 (2012)
2. Chen, J., Kellokumpu, V., Zhao, G., Pietikinen, M.: RLBP: robust local binary pattern. In: Proceedings of the British Machine Vision Conference (2013)
3. Sharma, G., Hussain, S., Jurie, F.: Local higher-order statistics (LHS) for texture categorization and facial analysis. Lecture Notes in Computer Science, vol. 7578. Springer, Heidelberg (2012)
4. Utrobin, V.A.: Physical interpretation of the elements of image algebra. Phys. Usp. **47**, 1017–1032 (2004)
5. Hafemann, L.G.: An analysis of deep neural networks for texture classification. Dissertation presented as partial requisite to obtain the Masters degree, Curitbia, Brazil (2014)

Method of Calculation of Upper Bound of Learning Rate for Neural Tuner to Control DC Drive

Anton I. Glushchenko[✉]

A.A. Ugarov Stary Oskol Technological Institute (Branch) NUST "MISIS", Stary Oskol, Russia
a.glushchenko@ieee.org

Abstract. A two-loop cascade control system of a DC drive is considered in this research. The task is to keep transients quality in both speed and armature current loops. It is solved by a usage of P- and PI-controller parameters neural tuner, which operates in real time and does not require a plant model. The tuner is trained online during its functioning in order to follow the plant parameters change, but usage of too high values of a learning rate may result in instability of the control system. So, the upper bound of the learning rate value calculation method is proposed. It is based on Lyapunov's second method application to estimate the system sustainability. It is applied to implement adaptive control of a mathematical model of a two-high rolling mill. Obtained results show that the proposed method is reliable. The tuner allowed to reduce the plant energy consumption by 1–2% comparing to conventional P-controller.

Keywords: Control system stability · Learning rate · DC drive · Neural tuner

1 Introduction

Adaptive control systems development is quite actual for industrial plants of high power nowadays [1]. The main idea is to compensate plant parameters change caused by switching to another functioning mode or equipment wearing by linear P/PI-controllers parameters adjustment. This allows to decrease energy consumption of the nonlinear plant. Electric DC drives of different industrial machines (for an instance, rolling mills) are of particular interest. Their control systems include two loops [2], each of which contains nonlinear plant: (1) an armature winding, (2) mechanics, which parameters drift due to mentioned reasons. Both loops are based on P- and PI-controllers, which should be adjusted. Most of methods to make such an adjustment [3–5] are based on a plant, a reference model or a state observer usage. Obtaining the plant model or the observer is a complicated task considering real production. The problem is to choose the adjustment step size value, using the reference model.

We have proposed a neural tuner [6, 7], which is able to solve the problem in question and does not need mentioned models. It combines expert systems and neural networks. A rule base defines moments when to train the network online and calculates the learning rate. But a method is needed to check whether this learning rate is not too high to make the control system unstable.

© Springer International Publishing AG 2018
B. Kryzhanovsky et al. (eds.), *Advances in Neural Computation, Machine Learning, and Cognitive Research*, Studies in Computational Intelligence 736,
DOI 10.1007/978-3-319-66604-4_16

2   Definition of the Tuner and Problem Statement

The DC drive is in operation at the moment. Its control system consists of two loops: for armature current (I_A) and speed (n). The current parameters of both speed and armature current controllers are known. They have been calculated for one of the functioning modes, but are not optimal for all other ones. The speed setpoint is changed like a ramp with known intensity. The required transient quality indexes (overshoot etc.) are also known for both loops. The mechanics and the armature winding change their parameters. The task is to keep required transient quality.

The neural tuners operate as follows. Each of them contains a rule base to (1) define moments when to train a neural network online, (2) calculate a learning rate. Rules for both tuners are shown in [6, 7]. The neural networks structures are calculated using the method described in [6]: for the armature current loop it is 5-14-2 (output neurons calculate proportional K_P and integral K_I controller gains), for the speed loop it is 2-7-1 (output neuron calculates K_P). The initial state for both these networks is set using extreme learning machine [8] allowing to form K_P, K_I values, which have been used before tuners installation, at networks outputs. Then the tuners are trained online using the backpropagation method [9]. The task is to calculate the upper bound for the output neurons learning rates $\eta_1^{(2)}$, $\eta_2^{(2)}$ to keep the control system stable. The learning rate for the hidden layer is constant (10^{-4}). Tuners are called every Δt seconds [10].

3   Stability of the Control System

Having analyzed researches on the matter of the systems with intelligent controllers and tuners development [11, 12], Lyapunov's second method is chosen for the system in question. The main state variable for both control loops is the error (e), which is the difference between the setpoint $r(t)$ and the plant (the mechanics, the armature winding) output $y(t)$. According to [9], the neuron calculating K_P is trained in proportion to the speed of e change: $e_1(t) = e(t) - e(t - \Delta t)$, whereas the neuron calculating K_I is trained in proportion to e value: $e_2(t) = e(t)$. So Lyapunov's function (1) is proposed.

$$V(E) = \frac{1}{2} \sum_{i=1}^{N_{ouput}} e_i^2(t). \tag{1}$$

N_{output} is the network outputs number. $V(E)$ derivative is calculated as (2) using (3).

$$\Delta V = \frac{1}{\Delta t} \sum_{i=1}^{N_{ouput}} \Delta e_i(t) \cdot e_i(t) + \frac{1}{2\Delta t} \sum_{i=1}^{N_{ouput}} \Delta e_i^2(t). \tag{2}$$

$$
\begin{cases}
e_1(t) & = \Delta e(t) = (r(t) - y(t)) - (r(t - \Delta t) - y(t - \Delta t)) \\
e_2(t) & = e(t) = r(t) - y(t) \\
\Delta e_1(t) & = r(t) - 2r(t - \Delta t) + r(t - 2\Delta t) - y(t) + 2y(t - \Delta t) - y(t - 2\Delta t) \\
\Delta e_2(t) & = r(t) - y(t) - r(t - \Delta t) + y(t - \Delta t).
\end{cases}
\tag{3}
$$

Let's substitute $e_i(t)$ and $\Delta e_i(t)$ in (2) with (3). The equation for the PI-controller is (4). Such equation has also been obtained for the P-controller. ΔV should be lower than zero to follow the sustainability sufficient condition. It can be calculated at every moment of the system functioning. This approach is used to calculate the learning rate upper bound $\eta^{(2)}_{k\,max}$ $(k = 1, 2)$.

$$
\begin{aligned}
\Delta V(t) &= [6y^2(t) + y(t)(-12r(t) + 14r(t - \Delta t) - 4r(t - 2\Delta t) - 14y(t - \Delta t) \\
&+ 4y(t - 2\Delta t)) + 9y^2(t - \Delta t) + y(t - \Delta t)(14r(t) - 18r(t - \Delta t) + 6r(t - 2\Delta t) \\
&- 6y(t - 2\Delta t)) + y^2(t - 2\Delta t) + y(t - 2\Delta t)(-4r(t) + 4r(t - \Delta t)) + 6r^2(t) + r(t) \\
&\cdot (-14r(t - \Delta t) + 4r(t - 2\Delta t)) + 9r^2(t - \Delta t) - 6r(t - \Delta t)r(t - 2\Delta t) + r^2(t - 2\Delta t)] \leq 0.
\end{aligned}
\tag{4}
$$

4 Upper Bound of Learning Rate Calculation

Let ΔV has been calculated for present moment of the system functioning for all control loops. If it is negative, then the control loop is stable and the controller parameters can be made both higher or lower. If it is not negative, then the controller parameters can be made lower only to increase the system stability. In both cases the desired plant output $y_{st}(t)$ value, where ΔV would be negative, should be calculated for the current moment. In order to do that, inequality (4) should be solved, considering $y(t)$ as a variable. If a range of y_{st} is found, then a certain y_{st} is chosen depending on the sign of ΔV, whether the previous setpoint value r_old is higher or lower than the current one $r(t)$ and whether K_P or K_I is going to be made higher or lower (see Fig. 1).

Let's consider how to calculate $\eta^{(2)}_{k\,max}$. The found y_{st} value allows to calculate changes of the control action $u(t)$ value, see the expression (5). Then maximal value of K_P and K_I correction can be calculated using (6) (dt is the controller functioning sampling time).

$$
\Delta u(t) = K_P \cdot \left(e(t) - e_{st}(t) \right) = K_P \cdot \left(r(t) - y(t) - r(t) + y_{st}(t) \right) = K_P \cdot \left(y_{st}(t) - y(t) \right).
\tag{5}
$$

$$
\begin{cases}
\Delta K_1 & = \Delta K_P \leq \Delta u(t)/e(t) \\
\Delta K_2 & = \Delta K_I \leq \Delta u(t)/(e(t) \cdot \Delta t/dt).
\end{cases}
\tag{6}
$$

Then, using the backpropagation method, the mathematical model and the fact that a linear activation function is used in the output layer of both tuners neural networks, the Eq. (7) is obtained for the learning rate upper bound calculation. N_{hidden} is the number of the neurons in the hidden layer. $o_j^{(1)}$ is the jth hidden neuron output. Then, if $|\eta^{(2)}_k| < |\eta^{(2)}_{k\,max}|$, then $\eta^{(2)}_k$ is used for training, otherwise $\eta^{(2)}_k = \eta^{(2)}_{k\,max}$.

$$\eta_{k\,\text{max}}^{(2)} = \frac{\Delta K_k}{e_k^{(2)} \left(\sum_{j=1}^{N_{hidden}} \left(\left(O_j^{(1)} \right)^2 \right) + 1 \right)}, \quad (k = \overline{1,\,2}).$$ (7)

5 Experimental Results

Experiments have been conducted using a two-high rolling mill 1000 mathematical model (Fig. 2). Initial values for controllers were calculated according to: (1) symmetrical optimum for the armature current loop, (2) technical optimum for the speed loop.

The speed setpoint (*setpoint* block in Fig. 2) was changed as follows: 0 rpm → 60 rpm → 0 → −60 rpm. The main requirement for the speed loop is to keep 0.35–0.55% overshoot σ. $K_{sf} = 0.637$, $K_{cf} = 9.407 \cdot 10^{-4}$, $K_A = 41.7$, $T_E = 0.036$ s, $T_M = 4798$ s, $K_P = 0.489$, $K_{P\,sp} = 1.745$, $K_I = 13.649$.

As for the armature current loop, the main requirement was to keep 3% overshoot σ_{curr}. There were two experiments. Only one tuner was used during each experiment.

As for the first one, the armature winding parameters drift was modeled. The armature winding gain K_A and time constant T_E were changed in the range 0.8–1.2 from their nominal values. The results are depicted in Fig. 3. Figure 3h shows a certain range of time from Fig. 3a to demonstrate several transients. All moments when $|\eta_1^{(2)}|$ and $|\eta_2^{(2)}|$ equal to nil were deleted from Fig. 3d and e. As a result, Fig. 3i and j were obtained. It can be concluded that almost at every moment absolute value of the learning rate upper

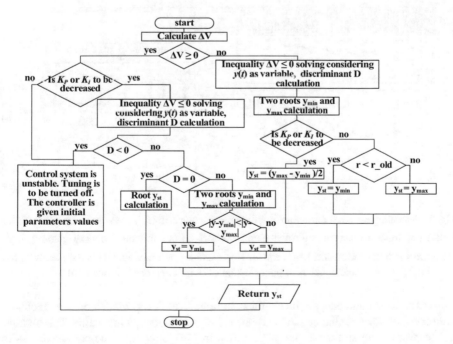

Fig. 1. Algorithm to calculate y_{st}

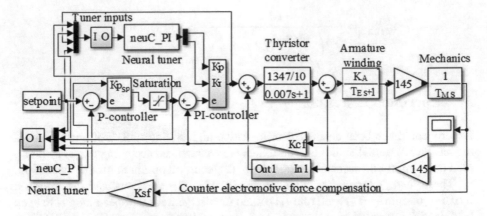

Fig. 2. Two-high rolling mill DC drive model (K_{sf} and K_{cf} are the speed and the armature current feedback parameter, K_A is the armature winding gain, T_E is the armature winding time constant, T_M is the mechanics time constant, K_P and K_I are the current controller parameters, $K_{P\ sp}$ is the speed controller parameter, e is error, *neuC_P* and *neuC_PI* are neural tuners.)

bound $|\eta^{(2)}_{k\,max}|$ was higher than the absolute value of the learning rate obtained from the triggered rule $|\eta^{(2)}_k|$ ($k = 1, 2$). When $|\eta^{(2)}_{k\,max}| < |\eta^{(2)}_k|$, $\eta^{(2)}$ was used as a learning rate. Comparing to an ordinary PI-controller, the tuner allowed to reduce the rolling mill energy consumption by 1.5%.

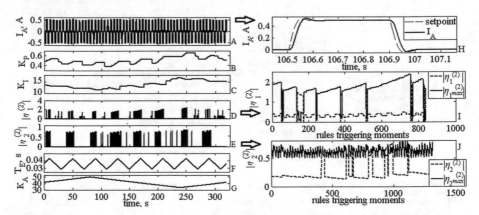

Fig. 3. Modeling results for armature winding parameters drift (I_A is the armature current, K_P and K_I are the armature current controller parameters, K_A is the armature winding gain, T_E is the armature winding time constant, $|\eta^{(2)}_1|$ and $|\eta^{(2)}_2|$ are absolute values of output neurons learning rate, $|\eta^{(2)}_{1max}|$ and $|\eta^{(2)}_{2max}|$ are absolute values of upper bound of output neurons learning rate)

As for the second one, the mechanics parameters drift was modeled. The mechanics time constant T_M was changed in the range 0.5–1.5 from its nominal value. The obtained results were quite similar to the ones shown in Fig. 3, so the same conclusion as for

Fig. 3 can be made. The tuner allowed to reduce the plant energy consumption by 2% comparing to conventional P-controller.

6   Conclusion

The method to calculate the upper bound of the learning rate for the neural tuner allowed to ensure the sufficient sustainability condition fulfillment for all conducted experiments. There were no cases of the algorithm branch execution, which was responsible for tuning stoppage. The DC drive energy consumption was reduced by 1.5–2%. Taking these into account, it can be concluded that proposed method is reliable.

Acknowledgments.   This work was supported by the Russian Foundation for Basic Research. Grant No 15-07-06092.

References

1. Astrom, K.J., Wittenmark, B.: Adaptive Control, 2nd edn. Dover, New York (2008)
2. Leonhard, W.: Control of Electrical Drives. Springer, Berlin (2001)
3. Son, Y.I., et al.: Robust cascade control of electric motor drives using dual reduced-order PI observer. IEEE Trans. Ind. Electron. **6**(62), 3672–3682 (2015)
4. Rotach, V.Y., Kuzishchin, V.F., Petrov, S.V.: Tuning of industrial controllers from the transient responses of control systems without approximating them by analytical expressions. Therm. Eng. **10**(57), 872–879 (2010). doi:10.1134/S0040601510100083
5. Liu, Y., et al.: Model reference adaptive control-based speed control of brushless DC motors with low-resolution Hall-effect sensors. IEEE Trans. Power Electron. **3**(29), 1514–1522 (2014). doi:10.1109/TPEL.2013.2262391
6. Glushchenko, A.I.: Neural tuner development method to adjust PI-controller parameters on-line. 2017 IEEE Conf. Russ. Young Res. Electr. Electron. Eng. (2017). doi:10.1109/EIConRus.2017.7910689
7. Eremenko, Y., Glushchenko, A., Petrov, V.: On PI-controller parameters adjustment for rolling mill drive current loop using neural tuner. Procedia Comput Sci **103**, 355–362 (2017). doi:10.1016/j.procs.2017.01.121
8. Tang, J., Deng, C., Huang, G.B.: Extreme learning machine for multilayer perceptron. IEEE Trans. Neural Netw. Learn. Syst. **4**(27), 809–821 (2016)
9. Demuth, H.B., et al.: Neural Network Design. Martin Hagan, USA (2014)
10. Eremenko, Y.I., Glushchenko, A.I., Fomin, A.V.: On development of method to calculate time delay values of neural network input signals to implement PI-controller parameters neural tuner. International Conference on Industrial Engineering. Applications and Manufacturing (ICIEAM), pp. 1–6. IEEE, Chelyabinsk (2016)
11. Changa, W.-D., Hwangb, R.-C., Hsieha, J.-G.: A self-tuning PID control for a class of nonlinear systems based on the Lyapunov approach. J. Process Control **12**, 233–242 (2002). doi:10.1016/S0959-1524(01)00041-5
12. Rossomando, F.G., Soria, C.M.: Design and implementation of adaptive neural PID for non linear dynamics in mobile robots. IEEE Lat. Am. Trans. **4**, 913–918 (2015)

Intelligent Diagnostics of Mechatronic System Components of Career Excavators in Operation

S.I. Malafeev[1(✉)], S.S. Malafeev[1], and Y.V. Tikhonov[2]

[1] Joint Power Co. Ltd., Moscow, Russia
sim@jpc.ru
[2] RPC Electrical Engineering: Research & Practice, Moscow, Russia
geovas333@gmail.com

Abstract. The article provides the results of application of artificial neural networks for diagnosis of the condition of electrical mining machinery as well as the description of data collection and processing of intelligent system structure and a condition of components of mechatronic systems analysis algorithms using neural networks. Information is provided on practical implementation of algorithms in information and diagnostic systems of career excavators developed by Joint Power Co. Ltd., Moscow.

Keywords: Diagnosis · Neural network · Mechatronics · Resource · Reliability · Simulation · Automation · Control · Electrical equipment · Efficiency · Converter · Controller

1 Introduction

The most important factor in the development of economy in modern conditions is to accelerate the solution of actual problems of increase of the efficiency of material, energy and information resources in various sectors of the economy. Modern technical means of measurement allow to control many process parameters during operation of machinery and equipment, to convert and transfer data, etc. [1, 2]. However, increasing the amount of data essentially complicates its analysis by a human. Therefore, it is important to establish automatic mining systems that provide continuous monitoring of the equipment during its life cycle, identifying deviations in the early stages of defects, forecasting of service life, the accumulation of information on the work and malfunction etc. [3].

In severe conditions all components of mining and metallurgical machinery are subject to intensive external influences, resulting in accelerated wear [4]. The immobilization time in case of failure of components is accompanied by great economic losses related mainly to underproduction. The repair and maintenance of excavators now accounts for 30% of the cost of mineral extraction; more than 20% of the downtime associated with the repairs, about 40% of unproductive ineffective work. The process of the repairs have low level of mechanization, more than 90% of repairs are performed without prior analysis of equipment defects using nondestructive testing methods.

© Springer International Publishing AG 2018
B. Kryzhanovsky et al. (eds.), *Advances in Neural Computation, Machine Learning, and Cognitive Research*, Studies in Computational Intelligence 736,
DOI 10.1007/978-3-319-66604-4_17

A study of emergency downtime of the excavators of "Kuzbassrazrezugol" has shown that the main causes of malfunction are mechanical failure (60%) and defects of electrical machinery (18%). Detailed failure analysis of career excavator is given in [5]. The increase in resource is a major reserve to save money, materials, energy and labor costs.

Operational control and continuous assessment of a residual resource significantly improve the efficiency of the equipment. In addition, failures of electrical and mechanical equipment may create risks to life and health of personnel. Continuous monitoring, intelligent diagnosis and evaluation of the resource allow for recovery and repair of equipment depending on its condition and reduce, thus, risks.

Organization of intellectual ventures in mining industry is a promising development trend. "Intellectual career" is a technology of future that implements robotic mining. Various companies are actively working on the use of new information technologies, intelligent control, monitoring and diagnostics on their machines. Since 2005, the overall technology of the P&H company (a division of Joy Global Inc.) is based on system Centurion [6] for managing and collecting data. Hardware and software components of this system are implemented using neural networks and specifically designed to create a highly efficient and closely linked network of data control, management and transfer.

The system allows to provide a high-tech access and control of drive system, to minimize the cycle time, to analyze the productivity of the excavator, to control operator's actions, as well as partially automate control of the excavator, which leads to significant increase in machine performance.

A promising direction in intelligent diagnostics of components of mining machines is the use of neural networks [7–9]. Problems of diagnostics are reduced to choice of informative parameters, organization of data collection system architecture, synthesis of network, definition of its parameters and its training.

In this paper are examined the results of research and simulation of intelligent diagnostic system of mechatronic complex of career excavator components.

2 The Organization of the Intelligent Diagnostics of Mechatronic Complex Components

Intelligent excavator is a machine with a high level of organization of management processes, monitoring and diagnostics, efficient human-to-machine and communication interfaces, adaptive to changes depending on the mining conditions and harmonious with the systems of power supplying, transportation and automated enterprise management [10].

For monitoring and subsequent analysis of work of excavator's mechatronic complex and its subsystems throughout the life cycle there was developed hardware and software complex "Electronic engineer", providing the following key functions:

- data collection from all components of the mechatronic complex;
- logging of the life cycle for mechatronic complex;
- data integrity checking;
- data analysis in real time using source-based data processing algorithms, including:

- evaluation of the efficiency of the driver, the transport and organization of mining works;
- estimation of condition and resource of components of excavator mechatronic complex;
- evaluation of the external environment (power lines, ambient temperature, power quality, condition of bottom, etc.)

- visualization of data on a local operator's panel as well as on an external device (e.g., computer) within Internet connection;
- communication of excavator mechatronic complex with machines and staff of mining enterprises and design organizations.

The scheme of algorithm of software operation of the system is shown in Fig. 1. Data logging is carried out directly from local information and diagnostic modules of excavator, the connection between them and APCS level computer is carried out according to the CAN Protocol.

The data may be stored in computer memory or on a dedicated server, and in the simplest case, the recording and storage of data may be performed using the recorder to a flash-card.

Checking the data integrity is carried out automatically at the network level of interaction of APCS and local subsystems, however, depending on the particular equipment, the possibility of data verification can be implemented on the application level.

The system "Electronic engineer" contains three subsystems of data processing which perform:

- assessment of the effectiveness of mechatronic complex of career excavator;
- assessment of equipment of mechatronic complex of excavator;
- evaluation of components resource of mechatronic complex of excavator.

3   Neural Network for Data Processing

Mechatronic system of career excavators consist of a standard set of components, power transformers, semiconductor converters (active rectifiers, inverters, etc.), AC or DC motors, switchgear, cables and controls.

Each component has its own properties regarding performance and occurrence of malfunctions and therefore requires its own set of state variables that characterize its work. Application of neural networks for diagnostic information processing uses current information about the device and provides parallel data processing.

Neural analyzer is built on the topology of "recurrent dual-layer perceptron" performing clustering of the input data.

There are two ways of clustering data, the first involves the assessment of mechatronic system as a whole, using 3 neurons in the output layer corresponding to normal, acceptable and emergency operating modes.

The second method involves a more in-depth data analysis, which first identifies a faulty component with a mechatronic complex, and then determining its condition, the

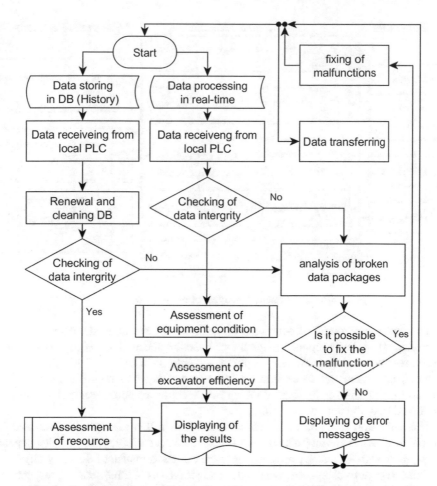

Fig. 1. Operation scheme of hardware-software complex "Electronic engineer"

number of neurons involved in the analysis process, is determined by the total number of possible states of the equipment.

Signals of neural analyzers of mechatronic complex components of the excavator go to the output neural analyzer (Fig. 2). The output neural analyzer evaluates the current state of the equipment and, in the case of malfunction, alerts.

Creating neural networks is performed in Matlab using "nftool". Then neural network is trained and then tested on a data set that is different from that with which the training has been carried out. It is necessary to exclude the existence of "memory effect".

Trained and tested neural network is saved in a separate file. This file is used in real-time simulation. Matlab also supports the ability to use built neural networks in third-party applications.

Let's examine in more detail the structure of a neural network, analyzing the state of twin-engine electric drive of career excavator.

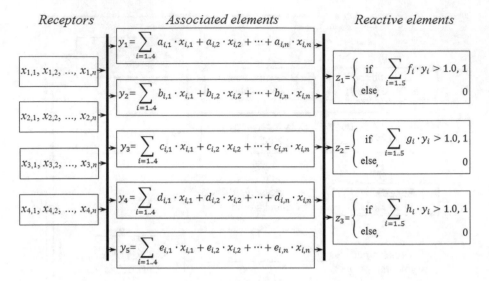

Receptors   *Associated elements*   *Reactive elements*

$x_{1,1}, x_{1,2}, ..., x_{1,n}$

$x_{2,1}, x_{2,2}, ..., x_{2,n}$

$x_{3,1}, x_{3,2}, ..., x_{3,n}$

$x_{4,1}, x_{4,2}, ..., x_{4,n}$

$$y_1 = \sum_{i=1..4} a_{i,1} \cdot x_{i,1} + a_{i,2} \cdot x_{i,2} + \cdots + a_{i,n} \cdot x_{i,n}$$

$$y_2 = \sum_{i=1..4} b_{i,1} \cdot x_{i,1} + b_{i,2} \cdot x_{i,2} + \cdots + b_{i,n} \cdot x_{i,n}$$

$$y_3 = \sum_{i=1..4} c_{i,1} \cdot x_{i,1} + c_{i,2} \cdot x_{i,2} + \cdots + c_{i,n} \cdot x_{i,n}$$

$$y_4 = \sum_{i=1..4} d_{i,1} \cdot x_{i,1} + d_{i,2} \cdot x_{i,2} + \cdots + d_{i,n} \cdot x_{i,n}$$

$$y_5 = \sum_{i=1..4} e_{i,1} \cdot x_{i,1} + e_{i,2} \cdot x_{i,2} + \cdots + e_{i,n} \cdot x_{i,n}$$

$$z_1 = \begin{cases} \text{if } \sum_{i=1..5} f_i \cdot y_i > 1.0, & 1 \\ \text{else,} & 0 \end{cases}$$

$$z_2 = \begin{cases} \text{if } \sum_{i=1..5} g_i \cdot y_i > 1.0, & 1 \\ \text{else,} & 0 \end{cases}$$

$$z_3 = \begin{cases} \text{if } \sum_{i=1..5} h_i \cdot y_i > 1.0, & 1 \\ \text{else,} & 0 \end{cases}$$

Fig. 2. Structure of neural analyzer of twin-engine electric drive

The inputs (receptors) of neural network receive the normalized values of currents and voltages (total of 4 inputs, input number is designated as i), hence $x_{i,j}$ is the state of the i-th receptor at time j (memory depth of perceptron – n).

Next, the signals go to an internal (associative) layer of neural network, where summation of signals is performed taking into account weighting coefficients $a_i \ldots e_i$, obtained during the network training.

The activation function of the associative layer in this example is linear and the internal feedback is absent (this, in the absence of the need to study the complex dependencies of the signals, can significantly improve the performance of the network).

External (reactive) layer of the neural network consists of three neurons, respectively for normal, acceptable and emergency modes. Activation functions of responsive neurons constitute relay elements. Weight coefficients of output layer in this case is $f_i \ldots h_i$.

The signals of local neural analyzers of career excavator mechatronic complex go to the output neural analyzer, its neural network structure is characterized that it uses an internal feedback. For the output neural analyzer the equation for y_m is as follows:

$$y_m = \sum_{i=1..N} a_{i,1}x_{i,1} + a_{i,2}x_{i,2} + \ldots + a_{i,n}x_{i,n} + \sum_{i=1..K} b_{i,1}y_{i,1} + b_{i,2}y_{i,2} + \ldots + b_{i,n}y_{i,n},$$

where $a_{i,j}$ and $b_{i,j}$ are a weight coefficients, N is a total number of receptors, K is a total number of hidden neurons, $x_{i,j}$ is a value on input of the i-th receptor at the time j, $y_{i,j}$ is a value at output of the i-th neuron of hidden layer at time j, n is a memory depth of perceptron.

During studies of neural analyzers varied the following parameters: voltage of power line feeding the excavator ($\pm 20\%$ of nominal), temperature of power transformer of drives and its windings (±40% of nominal), resistance of rotor windings of DC electric drives (20% of nominal).

Additionally, were simulated situations in which fired protection (current and temperature circuit breakers).

A total of 50 experiments were carried out. In 45 experiments, neural analyzer assessed adequately the condition of components of mechatronic complex (normal, overload or accident), 4 experiments presented false-positive failure result and 1 false-negative normal result.

4   Practical Implementation of Algorithms in the Diagnostic System

The developed algorithms of state estimation of components of mechatronic systems: circuit breakers [11], transformers [12], motors [13] used in information-diagnostic system of career excavator developed by the Joint Power Co. Ltd. [14].

Operational data are received from the information-diagnostic system and converted according to special algorithms for easy perception and efficient analysis of emergency situations, monitoring of equipment condition, parameters of power network.

Data is displayed on the monitor and stored on the server. Important information for operation evaluation is stored in software modules and processed to analyze the work efficiency of the excavator and to assess its reliability.

During the work is provided logging of: basic processes, changes of equipment condition, malfunctions protocols etc. Data is stored at the server and transmitted to the center. The retention period of records depends on type of process and type of equipment.

5   Conclusion

The increase in life of machines and equipment in the mining and metallurgical industries is achieved through continuous monitoring and forecasting during operation. Operational control and continuous assessment of a residual resource significantly improve the efficiency of the equipment.

Intelligent systems for diagnostics of electric drives and equipment provide minimizing cycle time, control of excavator productivity, effective protection, monitoring actions of the operator, and allow to partially automate the management process, which leads to increased productivity.

Information systems provide:

- full control of all main operating parameters, pressures, condition of equipment components;
- analysis and presentation in a convenient form of data on main technological parameters of equipment operation.

Monitoring data on excavator operation are transmitted via Internet to central server and accessible anywhere in the world.

References

1. Higgs, P.A., Parkin, R., Jackson, M., et al.: A survey on condition monitoring systems in industry. In: Proceedings of ESDA 2004: 7th Biennial ASME Conference Engineering Systems Design and Analysis, Manchester, UK, 16 p., 19–22 July 2004
2. Valavanis, K.P. (ed.): Applications of Intelligent Control to Engineering Systems, 423 p. Springer (2009). ISBN: 978-90-481-3017-1
3. Buse, D.P., Wu, Q.H.: IP network-based multi-agent systems for industrial automation. In: Information Management, Condition Monitoring and Control Systems for Industrial Automation, 187 p. Springer (2007). ISBN-13: 9781846286469
4. Raza, M.A., Frimpong, S.: Cable shovel stress & fatigue failure modelling – causes and solution strategies review. J. Powder Metall. Min. (2013). doi:10.4172/2168-9806.S1-003
5. Roy, S.K., Bhattacharyya, M.M., Naikan, V.N.A.: Transactions of the institution of mining and metallurgy. Sect. A Min. Technol. **110**(3), 163–171 (2001). doi:10.1179/mnt. 2001.110.3.163
6. Centurion. Electric Mining Shovel DCS800. Peak Services, 112 p. P&H Mining Equipment Inc., Milwaukee (2010)
7. Vachtsevanos, G., Lewis, F., Roemer, M.: Intelligent Fault Diagnosis and Prognosis for Engineering Systems, 454 p. John Willey & Sons. Inc. (2006). ISBN: 978-0-471-72999-0
8. Pat. US No.: 7 873 581 B2. Int. cl. G06F 15/18; G06G 7/00, 18 January 2011
9. Sun, F., Zhang, J., Tan, J.C., Yu, W. (eds.): Proceedings of the 5th International Symposium of Neural Networks, ISSN. Advances in Neural Networks, Part II, Beijing, China, 847 p., 24–28 September 2008
10. Malafeev, S.I., Tikhonov, Y.V.: Intellectualization of a career excavator. In: Reports of the XXIII International Scientific Symposium, Miner's week – 2015, Moscow, pp. 619–626, 26–30 January 2015
11. Pat. RU No. 2550337. Int cl. G01R 31/327. Date of Publication 10.05.2015. Bull. No. 13
12. Pat. RU No. 2559785. Int cl. G01R 31/00; H01F 41/12. Date of Publication 10.08.2015. Bull No. 22
13. Pat. RU No. 2536669. Int cl. G06G 7/63. Date of Publication 27.12.2014. Bull No. 36
14. Malafeev, S.I., Novgorodov, A.A.: Design and implementation of electric drives and control systems for mining excavators. Russ. Electr. Eng. **87**(10), 560–565 (2016)

Emotion Recognition in Sound

Anastasiya S. Popova(✉), Alexandr G. Rassadin,
and Alexander A. Ponomarenko

Higher School of Economics, National Research University,
Nizhniy Novgorod, Russian Federation
{aspopova_5, grassadin}@edu.hse.ru,
aponomarenko@hse.ru

Abstract. In this paper we consider the automatic emotions recognition problem, especially the case of digital audio signal processing. We consider and verify an straight forward approach in which the classification of a sound fragment is reduced to the problem of image recognition. The waveform and spectrogram are used as a visual representation of the image. The computational experiment was done based on Radvess open dataset including 8 different emotions: "neutral", "calm", "happy," "sad," "angry," "scared", "disgust", "surprised". Our best accuracy result 71% was produced by combination "melspectrogram + convolution neural network VGG-16".

Keywords: Deep learning · Classification · Convolutional neural networks · Audio recognition · Emotion recognition · Speech recognition

1 Introduction

Human emotion recognition in the flow of multimedia data is an actual and actively developed field of computer science. The emotion classification problem has great potential for use in many applied industries, such as robotics, tracking systems and other systems with interactive user interaction. Solving of this problem allows to receive users feedback in a natural way, it does not require any additional users actions, simplifying and accelerating the interaction between computer and a person.

2 Materials and Methods

The classification problem can be considered as a constructing task of a function $y: X \rightarrow Y$, where X- is the set of various descriptions of objects, Y - is the finite class set. Thus, there is an unknown target dependence mapping y, whose values are known only at the objects of the finite training sample $X_m = \{(x_1,y_1), \ldots, (x_m,y_m)\}$. It is required to construct an algorithm $A:X \rightarrow Y$, which is able to categorize an arbitrary object $x \in X$. For images the desired map y is $y:R^n \rightarrow Y$, where n is the total number of pixels of the image. In case of audio signal recognition n is the number of samples in the recognition window (Fig. 1).

B. Kryzhanovsky et al. (eds.), *Advances in Neural Computation, Machine Learning, and Cognitive Research*, Studies in Computational Intelligence 736,
DOI 10.1007/978-3-319-66604-4_18

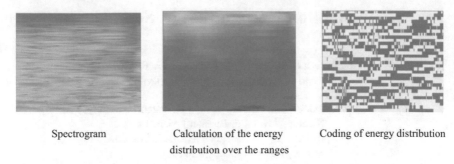

Spectrogram   Calculation of the energy   Coding of energy distribution
                     distribution over the ranges

Fig. 1. Illustration of the stages of J. Haitsma algorithm

There are two basic emotion theories: discrete theory [6], which is based on existence of universal basic emotions, they can differ in the number and types of basic emotions; the spatial theory assumes that emotions are decomposed into basis, thus the emotion can be represented as a point in the vector space [7, 8]. There are six basic emotions: neutral, angry, happy, sad and surprised. In this paper we adhere the discrete theory [6].

Previously, a number of methods have been proposed for classifying human emotions in audio, images [11, 12] or video sequence. Most of methods employ feature selection. That means those algorithms calculate the set of features which has vector representation (feature vector) and the classification is performed based on this vector.

For example, as a feature of image can be used a sign of presence of a smile on a face, the position and the shape of the mouth, the breadth of the eyes or the angle of the eyebrows In audio signal it is necessary to estimate the level of energy, the average level, the variance, the change in the height of the voice.

In this paper, we being inspired by the latest advances in computer vision and image recognition have set a goal to verify the approach in which classification of audio signal is reduced to the image recognition problem.

Nowadays, many problems related to sound processing have been successfully solved. There are many algorithms for working with sound files and many methods for its classification, which have various accuracy. Nevertheless, comparison of sound classification algorithms is subjectively, because experiments were conducted on different datasets and with different recognition problems.

The basic technique for processing audio signal is a fast Fourier transformation [5]. For example, the popular algorithms of J. Haitsma [3] and A. Wang [2] are both based on the analysis of time-frequency features obtained using Fourier window transformation. So, the first stage of these algorithms is preprocessing which builds a spectrogram of sound using a fast Fourier transform. Further, the J. Haitsma algorithm calculates the total energy in the subband for each time instant. The time distribution of energy is coded by function

$$F(n,m) = \begin{cases} 1 \text{ if } E(n,m) - E(n,m+1) - (E(n-1,m) - E(n-1,m+1)) > 0 \\ 0 \text{ if } E(n,m) - E(n,m+1) - (E(n-1,m) - E(n-1,m+1)) \leq 0 \end{cases}$$

where $E(n, m)$ – is energy of n frame in subrange m (Fig. 2).

| Spectrogram | Search for peaks | Combining peaks into pairs |

Fig. 2. illustrates the main stages of the algorithm A. Wang.

The algorithm A. Wang exploits another approach to recognition. It is based on the searching for the amplitude peaks of the spectrogram and matching them into pairs (constellations).

The main drawback of this algorithm is that it is rather complicated because peaks must be resistant to sound distortions. This complexity is well described in articles [6, 14]. Therefore, it is necessary to choose a large number of peaks throughout the entire area of the spectrogram. Each peak of the spectrogram in this algorithm is a point of the local maximum of energy. Usually, the number of peaks is determined for one frame. Using these limitations, it is possible to obtain peaks with the maximum probability of survival. Then the peaks are matching into pairs, so that each peak is connected to one or more peaks which are to the right of the time axis. This makes it possible to accelerate the algorithm by the following coefficient: $K \approx \frac{2^{(n_1 - n_2)}}{F^2}$, where n_1 is number of bits required to encode one peak, n_2 is number of bits required to encode pairs of peaks, F is branching factor.

This reduces the probability of collisions when hashing pairs of peaks. The probability can be approximately estimated as $p \approx [1 - 1 - pF]$, where p is the probability of survival of the spectrogram peak. As a result, the signal is specified by hash-pairs and codes of their displacements along the time axis.

The amount of memory used (in bits) to encode one pair estimation is:

$$n = \left] \log_2\left(\frac{F_s t_w}{2}\right) \right[+ \left] \log_2\left(\frac{\Delta t_a}{\Delta t_w}\right) \right[+]\log_2(2\Delta f_a t_w)[,$$

where F_s – frequency of signal sampling, is the size of the window used to build the spectrogram, Δt_w – a step of the window used to build the spectrogram, Δt_a, Δf_a – the maximum permissible distances along the time axis and the frequency axis between the peaks in the pair,][- rounding up.

Both algorithms performed well on the recognizing music tracks problem on the presented fragment with accuracy about 75% accuracy. Moreover, these methods of extracting features are used in the mobile application Yandex.Music [1] in the "Recognition" section and in the well-known mobile application "Shazam", which searches for music piece from the recorded fragment.

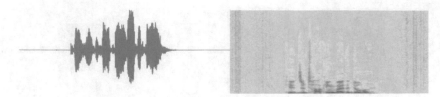

Fig. 3. On the left there is an oscillogram - a sequence of levels corresponding to the values of the voltage levels on the microphone membrane of the phrase «Kids are taking by the door» told by the actor with the emotion of happiness. On the right there is its spectrogram.

3 Examined Approach

Based on the fact that for today convolutional networks make it possible to get classifiers with accuracy of more than 99% for a large number of tasks and on different data sets, in this paper we examine the "straightforward" approach. It consists of reducing an audio classification problem to image recognition. There are many ways to represent the audio signal as a picture. In the simplest case, we can use an oscillogram (Fig. 3) directly as input image. Its explicitly depicts a sequence of values of the voltage levels sampled at identical small time intervals across the membrane of the microphone. In wav format this sequence of voltage levels is stored as a sequence of double-byte or three-byte integers corresponding to different 65,535 or 32 million values of voltage levels. However, if it is necessary to distinguish such signal characteristics as changing the pitch of the sound, the oscillogram is not a good visual representation of the audio signal. Therefore as a visual representation, we decide to use a spectrogram, which allows for experience musicians to see the structure of the music without addition processing.

Fig. 4. VGG-11 architecture

Fig. 5. VGG-16 architecture

As the training sample we took an open and labeled "RAVDESS" dataset [9], which includes records of 24 actors depicting 8 emotions: neutral, calm, happy, sad, angry, fearful, disgust" and "Surprised" (96 copies for "neutral", 120 for "surprised" and for 192 copies for the rest of emotions). Python, Numpy, Librosa were used as the basic tools for processing and analyzing sound files, Matplotlib was used to plot the graphs and was used for audios preprocessing.

The first stage of the experiment is the preprocessing of the audio file. At first, all the audio files were aligned by length. On this stage, by passing a sequence of membrane position levels to the input of the standard classifiers from the Sklearn library (Random Forsest, SVM, Adaboost), it is possible to achieve an accuracy of up to 30% with crossvalidation. From one point of view, the accuracy of 30% in this case is surprisingly large, and we were not expect that on such unprepared data, the classifiers will show accuracy more than a random choice corresponding to an accuracy of 12.5% for 8 classes.

At the next step, we scaled the signal by the volume; applied the lowpass and highpass filters to cut out the frequencies between 30 Zh and 2700 Zh because it is more suitable for human speech. Also we used the Voisssssce Activity Detection algorithm [17] to clean the voice. Then we applied the fast Fourier transform for each audio file and got spectrograms of sound. These spectrograms were used as images passing to the input to the image classifier. Here we used the VGG-11 convolutional neural network [4, 11] as the image classifier because it has relatively simple architecture and as a result has a fast speed rate. We used Keras library to construct the network architecture.

On the training stage, the classifier got an accuracy of about 98% on the training set and about 64% on the test set. The training set and test set did not overlap. It's were formed by choosing uniformly at random from the entire dataset and uniformly across all classes. 70% of the data set was used as a training set and 30% as a test set. The dependence of classification accuracy on the number of epochs in the learning process of the VGG-11 network with spectrograms is shown in Fig. 6a.

Changing network model from VGG11 to VGG16 and using melspectrograms [13, 14] (Fig. 4) instead of spectrograms (Fig. 5) gave us 71% accuracy on the test set and nearly 100% on training set. Mel scale is a perceptual scale of pitches judged by listeners to be equal in distance from one another, so mel and Hertz depends like logarithmic function:

$$m = 2595 \, \log_{10}\left(1 + \frac{h}{700}\right) = 1127 \, \ln\left(1 + \frac{h}{700}\right),$$

where m is mel and h is hertz. That is why it is more suitable for our problem. The dependence of classification accuracy on the number of epochs in the learning process of the VGG-16 network with melspectrograms is shown in Fig. 6b.

The confusion matrixes (Table 1) illustrate the errors between different classes. The rows of the table correspond to the correct classes and the columns correspond to the results of our model. Surprisingly that classification of a neutral emotion has small error. The model has done only eight mistakes with a calm emotion, which is very similar to the neutral, Unfortunately the model has some difficulties to separate happy and angry emotions. Most likely the reason for this is that they are the most strongest emotion and as a result their spectrograms are slightly similar, for example, both has many red color.

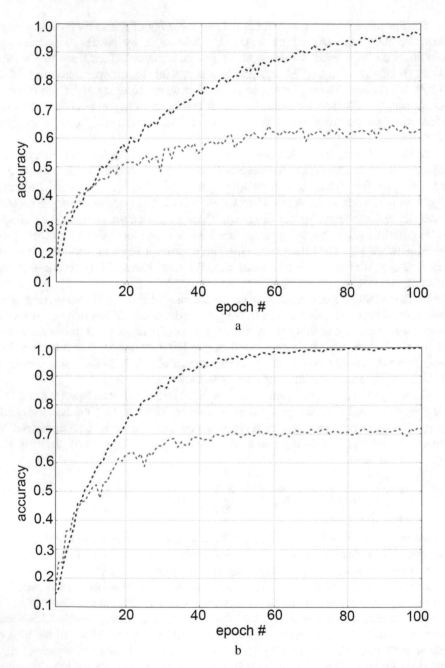

Fig. 6. Dependence of classification accuracy on the number of epochs in the learning process of the (a) VGG-11 network with spectrograms and (b) VGG-16 network with melspectrograms.

Table 1. Confusion matrix.

	Neutral	Calm	Happy	Sad	Angry	Fearfull	Disgust	Surprised
Neutral	21	8	0	0	0	0	0	0
Calm	7	46	1	1	0	1	2	0
Happy	0	1	26	7	6	9	5	4
Sad	0	2	10	31	2	3	9	1
Angry	0	1	1	0	43	2	5	6
Fearfull	0	1	3	2	6	34	6	6
Disgust	0	0	0	3	2	3	49	1
Surprised	0	0	2	1	1	8	12	12

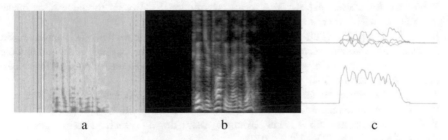

a   b   c

Fig. 7. (a) Spectrogram, (b) melspectrogram, (c) m Melspectrogram coefficients

4 Conclusions and Directions for Further Work

In this paper we proposed and verified an approach for classifying of human emotions in a sound fragment. A numerical experiment was performed using a convolutional neural network VGG-16 on preprocessed data. This experiment has shown not bad result - classification accuracy of 71% instead of 12.5% accuracy for random choice. This can be considered as a good result for the algorithm which does not use the extraction of complex audio-specific features which emotions of a particular type has.

In future we plan to use melspectrogram coefficients [15, 16] (Fig. 7).

Acknowledgments. The article was prepared within the framework of the Academic Fund Program at the National Research University Higher School of Economics (HSE) in 2017 (Grant №17-05-0007) and by the Russian Academic Excellence Project "5-100".

References

1. Krofto, E.: Kak YAndeks raspoznaet muzyku s mikrofona. In: Yet another Conference 2013, Moscow (2013). (in Russian)
2. Wang, A.: An industrial strength audio search algorithm. In: ISMIR, vol. 2003, pp. 7–13 (2003)
3. Haitsma, J., Kalker, T.: A highly robust audio fingerprinting system with an efficient search strategy. J. New Music Res. **32**(2), 211–221 (2003)

4. Choi, K., Fazekas, G., Sandler, M.: Automatic tagging using deep convolutional neural networks. arXiv preprint arXiv:1606.00298 (2016)
5. Cooley, J.W., Tukey, J.W.: An algorithm for the machine calculation of complex Fourier series. Math. Comput. **19**(90), 297–301 (1965)
6. Ortony, A., Turner, T.J.: What's basic about basic emotions? Psychol. Rev. **97**(3), 315 (1990)
7. Scherer, K.R.: What are emotions? And how can they be measured? Soc. Sci. Inf. **44**(4), 695–729 (2005)
8. Russell, J.A., Ward, L.M., Pratt, G.: Affective quality attributed to environments: a factor analytic study. Environ. Behav. **13**(3), 259–288 (1981)
9. Livingstone, S.R., Peck, K., Russo, F.A.: Ravdess: the Ryerson audio-visual database of emotional speech and song. In: Annual Meeting of the Canadian Society for Brain, Behaviour and Cognitive Science (2012)
10. Simonyan, K., Zisserman, A.: Very deep convolutional networks for large-scale image recognition. arXiv preprint arXiv:1409.1556 (2014)
11. Busso, C., Deng, Z., Yildirim, S., Bulut, M., Lee, C.M., Kazemzadeh, A., Narayanan, S.: Analysis of emotion recognition using facial expressions, speech and multimodal information. In: Proceedings of the 6th International Conference on Multimodal Interfaces, pp. 205–211. ACM (2004)
12. Zhang, Z.: Feature-based facial expression recognition: sensitivity analysis and experiments with a multilayer perceptron. Int. J. Pattern Recognit. Artif. Intell. **13**(06), 893–911 (1999)
13. Tsai, T.J., Morgan, N.: Longer features: they do a speech detector good. In: INTERSPEECH, pp. 1356–1359 (2012)
14. Eyben, F., Böck, S., Schuller, B.W., Graves, A.: Universal onset detection with bidirectional long short-term memory neural networks. In: ISMIR, pp. 589–594 (2010)
15. Ramachandran, A., Vasudevan, S., Naganathan, V.: Deep learning for music era classification. http://varshaan.github.io/Media/ml_report.pdf. Accessed 23 June 2017
16. Ishaq, M.: Voice activity detection and garbage modelling for a mobile automatic speech recognition application. https://aaltodoc.aalto.fi/handle/123456789/24702. Accessed 23 June 2017

The Classification of Objects Based on a Model of Perception

V.N. Shats$^{(\boxtimes)}$

St. Petersburg, Russia
vlnash@mail.ru

Abstract. This paper proposes a model of perception that allows animals to classify objects in the environment. We consider the transformation of semantic information in the four blocks of the model that imitate the mechanism of operation of sensory systems. Receptors convert external influences into stimuli that are transformed into sensations in accordance with the law of requisite variety via the data randomizing. Perception is formed by the generalization of all sensations, as well as the corresponding information accumulated by the animal. To find appropriate prototypes of objects forming classes, there is a compression of the processed information. This process is modeled via the creation of granules containing objects with close values for every feature. Granulation allows us to find the most probable class of the object corresponding to the average frequency value of its features. Algorithms for object classification on basis of the model and the invariant paradigm are identical and posses the simplicity and versatility at a high accuracy of the solution.

Keywords: Classification · Information granules · Perception model · Agent

1 Introduction

The existence of an animal requires the operation of its sensory systems, which continuously process the signals of the external world and form the animal's view of the environment. With this knowledge, the animal can classify objects in the environment to take timely behavioral decisions. There are many examples where an animal solves such problems much more efficiently than the most advanced technical systems that rely on modern methods of machine learning.

Currently, extensive research is being conducted on the processing of information in the animal's nervous system at the molecular and cellular levels [1]. The first direction of research is based on the Hodgkin–Huxley model, describing the physical and chemical processes in neurons. The other one is built on the McCulloch-Pitts formal neuron model, which reflects the structure of the nervous system. Note that in all studies, processes of nervous system interaction are considered outside any connection with the content of processed information.

This paper is devoted to the development of a perception model that is designed to solve a typical classification problem. Here, the particularity noted above plays an important role. The processes arising in nervous system are generated by the impact of objects from the external world on the receptors of numerous animal sensory systems,

© Springer International Publishing AG 2018
B. Kryzhanovsky et al. (eds.), *Advances in Neural Computation, Machine Learning, and Cognitive Research*, Studies in Computational Intelligence 736,
DOI 10.1007/978-3-319-66604-4_19

which are transformed into nerve impulses. After multiple transformations in neuronal circuits, they enter the brain, where at first created sensations, and then perception as an integral image of the environment [2, 3].

For an animal, it is important to distinguish those states of the environment that have the same properties in relation to its vital activity and correspond to a certain behavioral reaction [4]. The animal compares its perception of new impacts from the environmental with pre-existing options to choose a behavioral solution. From the standpoint of cognitive psychology, the animal performs of a prototype search [5, 6].

Therefore, the information processing is considered herein at a "rough" macroscopic, level rather than at the molecular or cellular level. The process model consists of several blocks with different functions of intelligent agents. To determine these functions, an approach was used that is common in physics: "guess the patterns" of processes, by comparing the known research data [7].

2   Theoretical Problems

2.1   Statement of the Problem

The purpose of this study is to develop a model of an animal's perception that is capable of solving the following classification problem:

Let $X = \{x_s | s \in (1, Ms)\}$ be the set of real objects consisting of the training samples $(1 \leq s \leq t)$ and the control samples $(t + 1 \leq s \leq Ms)$ described by the feature vectors $q^s = (q_1^s, \ldots, q_{Mk}^s)^T$, Q be the data matrix of the combined sample, and ω_i be the list of the numbers of the training sample objects of class $i \in (1, Mi)$. Assume that the features may be quantitative, nominal, ordinal, or mixed types; there are no missing data; the classes are disjoint. The goal is to find a rule for classifying the objects of the training sample and to assess its applicability to the objects of the control sample.

We assume that each feature is perceived by a certain sensory system, its receptors transform the feature q_k^s into a stimulus q_k^s. Then, the analogue of the data vector q^s is the stimulus vector q^s. The training sample corresponds to the accumulated experience of the animal, and the objects of the control sample correspond to the new environmental objects, which are subject to classification.

Consider the model of the information perception process, which consists of four blocks in which the following transformations are consistently performed: the evaluation of stimuli, their transformation into sensations, the compression of information and the definition of an object's class.

2.2   Initial Stages of Information Transformation

The evaluation block describes the operation of receptors (receptor fields) by the transformation of objects x_s into stimulus vectors q^s, the value of which is described in the corresponding latent language. It is shown below that the solution of the problem is determined by the frequency of features. Because these frequencies do not depend on the method used to identify the values of features (in absolute value, number or using any other scale), we can convert their reference frame to use any type of features, including mixed features.

In particular, for a non-quantitative feature, we should establish a partial-order relationship based on the set of options for its values under the arbitrary rule of their numbering. The changed value q_k^s will be equal to the number of corresponding options for the sequence numbers $1, 2, \ldots$. Let us assume that any such transformations of non-quantitative features are already incorporated into the matrix Q.

In the second block, the stimulus vectors $q^s = (q_1^s, \ldots, q_{Mk}^s)^T$ are transformed into vectors of sensations $c^s = (c_1^s, \ldots, c_{Mk}^s)^T$, where c_k is the sensation that reflects the individual properties of objects and phenomena, that affect the receptors of the kth sensory system. Here, the nervous system provides some mapping $R : q^s \to c^s$, under which the matrix $Q \to C$.

The physical realization of this mapping occurs by transmitting information along the nerve pathways from receptors to brain. It is accompanied by numerous transformations of information, which acquire a random character already when the physical stimulus is transformed into an electrical signal due to the peculiarities of threshold sensitivity of receptors. Therefore, the relationship between the magnitude of stimulus and sensation, which is measured on a latent scale, can only be approximate.

The streams of information reaching the brain, «with a minimum degree of accuracy» correspond to the initial information [10]. The sensory system can consider all of the errors arising from this transformation at the subsequent stages of perception.

At the same time, according to the law of requisite variety, the nervous system must be able to differentiate as many variants of the state of the environment as possible [8]. If each object will create unrepeatable sensations, then this requirement will be fulfilled. According to the above considerations, we assume that the function R leads to a randomization for which $c_k^s = q_k^s + \alpha v_s$ at all k and s, where v is a random variable uniformly distributed on the interval $[0, 1]$ and $\alpha \geq 0$ is a constant [9].

2.3 Compression of Information by Granulation

Further processing of sensory information is characterized by greater complexity of ongoing processes involved in creating a perception. It reflects the image of objects in the environment as a whole on the basis all sensations and all the sensory information already accumulated by the animals. In different parts of the brain, neurons are grouped into internally integrated ensembles of local neural networks from various sensory systems.

To ensure the storage of information and accelerate behavioral solutions, the information is compressed. The compression is accompanied by the elimination of redundant neurons and neural connections. As a result, a picture of the excited neurons arises that corresponds to the image of a particular object. The brain correlates this image with a collection of images that are stored in memory and the organism has previously met, and it decides which class the object corresponds to.

The idea of information compression is implemented in the third block of the model by structuring information on sensations. For this purpose, we divide the multidimensional space of sensations into Mk is dimensional parallelepipeds with the same relative edge length for each sensation. Then, we proceed to measure sensations in integers, called sensation levels, and we will be able to estimate the frequencies of objects falling into each of these parallelepipeds.

We introduce the vector of sensation levels $\boldsymbol{d}^s = \{d_k^s = m | 1 \le m \le n\}$, where $n > 1$ is an integer. To calculate d_k^s, we arrange the objects s in the order of non-decreasing values of c_k^s, and divide the range of values into n equal intervals. All objects falling into the interval $[m, m+1)$ have $d_k^s = m$. As a result, we find the matrix of sensation levels $\boldsymbol{D}$ of size $Ms \times n$.

Let $Z = Z_{k,m}$ denote the set whose elements for each k are lists of the objects $Z_{k,m}$ for which $d_k^s = m$. These elements group objects belonging to the ordered pairs $\{k, m\}$ and play the role of data granules. We introduce the analogous notation $\tilde{Z}$ and $\tilde{Z}_{k,m}$ for the objects of the training sample $s \in (1, t)$. We partition the lists $\tilde{Z}_{k,m}$ into lists of objects of individual classes $\tilde{Z}_{k,m}^i$, which we call information granules of class i.

Consider properties of the information granule $\tilde{Z}_{k,m}^i$. For any k, the sum $\sum_{m=1}^n l_{k,m}^i = N_i$, where $l_{k,m}^i$ is the length of granule and N_i is the number of objects class i in the training sample. Since some intervals may be "empty", then $0 \le l_{k,m}^i \le N_i$. Then, the frequency of ordered pairs k, m for any object $s \in \omega_i$ is $g_{k,m}^i = l_{k,m}^i / N_i$. Thus, data compression is accomplished by granulation, which allows separate objects to have frequency features $g_{k,m}^i$.

Let us pay attention to one more important feature of granulation. From the above dependences it follows that the value of m depends on the entire set of features values and is found with a random error. Therefore, the range of corresponding stimulus values q_k will have fuzzy, soft boundaries that will move along with the changes of n, combining different objects into a single granule. Nevertheless, we can assume that the granulation divides the objects according to the semantic content of the information. For example, objects could be grouped by a range of weight values or by color.

2.4   Classification of Objects as a Prototype Search

The effect of granulation manifests itself in the last block, since object classes can be determined using the simplest formulas of total probability. Observe that the estimate of probability $p(s \in \omega_i | d_k^s = m) = g_{k,m}^i$. For objects of the list ω_i, the event of occurrence of the vector $\boldsymbol{d}^s$ consists of a complete group of independent events $d_1^s, \ldots, d_{Mk}^s$. Therefore, the evaluation of probability of this event is

$$p(s \in \omega_i | \boldsymbol{d} = \boldsymbol{d}^s) = \frac{1}{Mk} \sum_{k=1}^{Mk} p(s = \omega_i | d_k^s = m) \tag{1}$$

Obviously, this estimate is the average over all k the frequency $g_{k,m}^i$, which depends on i. Its maximum determines the calculated class of an object s:

$$I(s) = \arg \max_{1 \le i \le Mi} p(s \in \omega_i | \boldsymbol{d} = \boldsymbol{d}^s) \tag{2}$$

Formulas (1) and (2) are obtained on the basis of probability theory. However, these same dependences follow from the theory of prototypes of cognitive psychology, according to which the object class is determined by the frequency of the most frequent combinations of features [5]. We can assume that these formulas determine the class of an object, not as a result of calculations, but by comparing the variants of images that differ in the average frequency of the features. Such an interpretation is consistent with modern ideas about the mechanisms of brain.

On the other hand, these formulas take into account cases where the probability density may be continuous for individual features of objects, but the density of objects and the conditional probability densities of features may have discontinuities. Therefore, unpredictable errors that arise when using Bayesian formulas in other methods, which are also based on probability theory, are excluded here.

According to (2), the class that is assigned to the object depends on the frequency $g^i_{k,m}$, which serves as an objective measure of the belonging of an object from the set X to a certain class pertaining to each sensation. Therefore, to classify the objects of the control sample, we need to use the values $g^i_{k,m}$ for the corresponding pairs of the training sample objects.

The set $\left\{g^i_{k,m}\right\}$ reflects the accumulated experience of the animal, which jointly characterizes the learning sample and the ω_i lists. This set represents the memory of the model in a condensed form. Note that $g^i_{k,m}$ gives a point estimate of the corresponding probability. However, its reliability depends on the interval estimation, which, in turn, depends on the length of sample.

3 Properties of the Model

It is important to note that the entire set of the above dependencies fully corresponds to the algorithm, regardless of object classification paradigm used [11, 12], and differs only in interpretation. Thus, the proposed model of perception is an instrument for the solution to problem of classification. However, we can use the results of research on the application of new paradigm.

The ratio of $I(s)$ and i^s determines the accuracy of training for the objects of the training sample and the accuracy of classification for the objects of the control sample. These characteristics depend on the design parameters α and n. The issues of verification and the selection of design parameters to ensure high accuracy of the solution, the influence of the relative sizes of the training and control samples, and the results of solving problems from the UCI repository are discussed in detail in [11].

Therefore, we confine ourselves to proving that training can be error-free for any data, because this result contradicts the established ideas about the possibilities of training.

Consider the granule $\tilde{Z}^i_{k,m}$, which contains the object s for some k and i. Let the object w also belong to this granule. Then, the following inequality holds $0 \leq \left|c^w_k - c^s_k\right| \leq \left(c^{\max}_k - c^{\min}_k\right)/(n-1)$, where $c^{\max}_k$ and $c^{\min}_k$ are, respectively, the maximum and minimum sensation values c_k. As, $n \to \infty$ the above parallelepipeds are tightened to a point. From this relation, it follows that $c^w_k = c^s_k$. In addition, the matrix

D becomes a sparse matrix, because objects can fall into no more than in t from an infinite number of intervals.

As noted above, if $\alpha > 0$, then all values c_k of will be different. Then, $w = s$ and $g_{k,m}^i = 1$ for all k. Hence it follows that the learning will be without error, since I $(s) = i^s$. If $\alpha = 0$, then the granules $\tilde{Z}_{k,m}^i$ may include an object whose number w is randomly changed for each k. Here, any relation between the average values $g_{k,m}^i$ of objects w and s is possible. Therefore, there is a risk of error. To eliminate the error, it is sufficient to take $\alpha > 0$.

4   Conclusion

In this paper, a model of perception is proposed. An application of the model provides a solution to the problem of classification. The model describes the processes in sensory systems to find a prototype of an object class in the accumulated information of an animal regarding descriptions of objects by their features.

When processing information in the model, three principles are consistently used:

(1) Maximum increase in the variability of data.
(2) Compression of information by splitting it into information granules.
(3) Separation of objects into classes according to the mean frequency of features of prototypes.

It is established that the developed model and the invariant paradigm of solving the classification problem use a single algorithm for computations. This conclusion points to the biological roots of the paradigm and indicates the plausibility of model. It gives the key to understanding the reasons for the universality and simplicity of the algorithm and the high accuracy of the results based on this paradigm.

References

1. Red'ko, V.G.: Evolution, neural networks, intelligence: models and concepts of evolutionary cybernetics, p. 220. URSS, Moscow (2015)
2. Pokrovsky, V.M., Korotko, G.F.: (eds.) Human physiology. In: Medicine, vol. 2, p. 368 (1997)
3. Smith, C.U.M.: Biology of Sensory Systems, p. 583. Aston University, Birmingham (2009)
4. Loskutov, A., Mikhaylov, A.S.: Fundamentals of the theory of complex systems, p. 620. Institute of Computer Studies, Moskow (2007)
5. Solso, R.: Cognitive Psychology, 6th edn, p. 589. Allyn and Bacon, Boston (2006)
6. Bratus, B.S.: General psychology. In: Gusev, A.N. (ed.) Sensation and Perception, vol. 2, p. 416. Academy, Moscow (2007)
7. Feynman, R.: The character of physical law. A series of lectures recorded by the BBC at Cornell University USA. Cox and Wyman, London (1965)
8. Zadeh, L.: Fuzzy sets and information granularity. In: Gupta, N., Ragadem, R., Yager, R. (eds.) Advances in Fuzzy Set Theory and Applications, pp. 3–18. North-Holland, Amsterdam (1979)

9. Ashby, W.R.: An Introduction to Cybernetics, 2nd edn, p. 295. Chapman and Hall, London (1957)
10. Granichin, O.N., Polyak, B.T.: Randomized Algorithms of an Estimation and Optimization Under Almost Arbitrary Noises, p. 209. Nauka, Moscow (2003)
11. Shats, V. N.: On new computing technology in machine learning. In: 17th Proceeding International Conference Neuroinformatics, vol. 2, pp. 148–158. MEPhI, Moskow (2015)
12. Shats, V.N.: Invariants of matrix data in the classification problem. Stochastic Optimization Informatics. **12**(2), 17–32 (2016). SPbGU, St. Petersburg http://www.math.spbu.ru/user/gran/optstoch.htm

An Approach to Use Convolutional Neural Network Features in Eye-Brain-Computer-Interface

A.G. Trofimov[1]([⊠]), B.M. Velichkovskiy[2], and S.L. Shishkin[2]

[1] National Research Nuclear University «MEPhI», Moscow, Russia
atrofimov@list.ru
[2] National Research Center «Kurchatov Institute», Moscow, Russia

Abstract. We propose an approach to use the features formed by a convolutional neural network (CNN) trained on big data for classification of electroencephalograms (EEG) in the eye-brain-computer interface (EBCI) working on short (500 ms) gaze fixations. The multidimensional EEG signals are represented as 3D-images that makes possible to apply them to CNN input. The features for EEG classifier are selected from first fully connected CNN layer's outputs. It is shown that most of them are useless for classification but at the same time, there were a relatively small number of CNN-features with a good separating ability. Their use together with the EEG amplitude features improved the sensitivity of a linear binary classifier applied to an EEG dataset obtained in an EBCI experiment (when participants played a specially designed game *EyeLines*) by more than 30% at a fixed specificity of 90%. The obtained result demonstrates the efficiency of the features formed by the CNN trained on big data even with respect to the essentially different classification task.

Keywords: Convolutional neural network · *AlexNet* · EEG · Eye-brain-computer interface · Binary classification

1 Introduction

The eye-brain-computer interface (EBCI) is a communication system in which brain bio-potentials (EEG) and eye-tracking data are jointly used to control technical devices. In the EBCI systems control is actually performed by means of a gaze, while the EEG data are used to solve the so-called "Midas touch problem" [1]. The essence of this problem is that spontaneous gaze fixation or gaze gestures cannot be suppressed by the user and can lead to issuing control commands even in the absence of any intention.

The EBCI working on short (500 ms) gaze fixations was proposed in [2]. The interface was tested using data collected when participants played a specially designed game *EyeLines*. In this game, a move was made by "clicking" on the button, selecting the ball in one of the cells of the game board and moving it to a free cell, all by means of gaze fixations only.

The principal part of any BCI is EEG feature extraction and algorithms for their classification [3]. The key problem remains the choice of features for the classifier,

© Springer International Publishing AG 2018
B. Kryzhanovsky et al. (eds.), *Advances in Neural Computation, Machine Learning, and Cognitive Research*, Studies in Computational Intelligence 736,
DOI 10.1007/978-3-319-66604-4_20

which provide the required class separability: in other words, there is no answer on how to select best features for a particular BCI.

A promising area in the feature engineering for classification is the automatic generation of features by means of learning on experimental data without the teacher (so-called feature learning). One of the models implementing this approach is the *Convolutional Neural Networks (CNN)* [4].

Convolutional neural networks have shown their effectiveness in solving classification problems in various applications including detection objects in images, speech and text recognition [5, 6]. Convolutional neural networks were also applied to the classification of EEG signals [7–9]. In this work, an approach to use the features generated by a CNN trained on big data is proposed to classify the EEG data in EBCI.

2   Problem Statement

We consider a set of L-channel EEG data which consists of N signals with the same number of time samples T divided into two classes: target and non-target. Target signals are related to the mental intention of the subject to perform the action and non-target signals are intention-free. Each EEG signal is represented as a matrix of dimension $L*T$.

To evaluate the quality of binary classification we will use the sensitivity (i.e. true positive rate) corresponding to the fixed (for example, 90%) specificity (i.e. true negative rate). High requirement for the classifier's specificity is motivated by a typically observed lower tolerance of the users to the EBCI false activations (false positives) than to the misses of response (false negatives, i.e. interface inactivity under the presence of mental intention to perform the action).

For successful classification of certain data, effective algorithms should be chosen or designed for creating feature vectors and for classifier training. As the classifier, we will use a shrinkage LDA classifier [10]. This type of classifier demonstrated best results among the linear classifiers with the feature vector constructed from the EEG amplitudes smoothed in time with a fixed time step and window. The results of the use of this classifier for the EBCI data were presented in [2].

In this paper, we aimed to improve the sensitivity of the linear shrinkage classifier by using the EEG features formed by a convolutional neural network.

3   Representation of the EEG Signals as Images

The use of convolutional neural networks for classification is possible in one of three ways: building a new model, retraining the already trained model and using network features with an arbitrary classifier. Since the available training sample size in BCI tasks are typically very small, it was decided to use a fully connected layer's outputs of the CNN trained on big data followed by training of a linear classifier that has higher robustness compared to a CNN classifier.

As the trained CNN, the *AlexNet* network was chosen [11]. This network is trained on the images from the *ImageNet* database used in the *Large-Scale Visual Recognition*

Challenge (ILSVRC) [12] containing over a million images assigned to one of the 1000 classes. Thus, the trained network is able to construct a rich representation for a wide range of images.

The network consists of 25 layers, including 5 convolution layers of different depths, layers of *ReLU*-neurons, as well as normalization, pooling and dropout layers, two fully connected layers and the network's output layer. The size of the network's input image is 227*227*3 (3 is the number of color channels). The dimension of the fully connected layers is 4096.

To use the *AlexNet* network it is necessary to convert the available raw data to the input network format, i.e. to represent them as images of dimension 227*227*3. Many methods for encoding time series as an image can be proposed [13]. In this paper, the EEG signal X of dimension $L*T$ is encoded to the required image format using the following algorithm.

1. Preprocessing of the EEG signals

To eliminate artifacts related to eye movement, fragments of the EEG data (sampling frequency of 500 Hz) starting 200 ms after the gaze fixation's start were only considered. The right border of the fragment was selected at 500 ms. Each fragment was corrected to the baseline calculated in the interval (200; 300) ms. A simple moving average smoothing with a 50 ms window and 20 ms time step was then applied, thus, the preprocessed EEG signal consisted of average amplitudes in the time windows 200–250, 220–270,…, 440–490 ms (total $T = 13$). The data used in this work were collected in the Department of Neurocognitive Technologies, National Research Center "Kurchatov Institute" (Moscow, Russia) in the experiment described in [2].

2. Reordering the EEG channels

The CNN trained on image data assume the two-dimensional convolution, so it is desirable that adjacent pixels of the input image correspond to adjacent EEG amplitudes both in time and space, i.e. through the channels. In this case, the result of convolution is a weighted average of EEG amplitudes in a compact space-time domain. Temporal adjacency is obtained in a natural way while the spatial adjacency requires reordering of the EEG channels, so that adjacent EEG channels correspond to adjacent pixels in the image.

Since only $L = 13$ fixed channels located in the parietal and occipital areas were used in this study, the reordering was carried out manually. The following order of the channels was defined:

PO7, P3, PO3, O1, Oz, POz, P1, Pz, P2, PO4, O2, PO8, P4.

In the case of larger number of channels, efficient algorithms can be proposed for mapping 2D-data to one-dimensional space (for example, based on Kohonen self-organizing maps [14]).

3. Resizing the $L*T$ matrix to the required dimension using bicubic interpolation.

4. Duplicating the image into three color channels.

4   Quality of Features Generated by the Convolutional Neural Network

The CNN features were composed from the outputs of the network's first fully con-nected layer of dimension 4096. Using data from one participant, the ROC curve for the threshold binary classifier applied separately to each CNN feature was obtained and the AUC (Area Under Curve) was calculated. It was found that AUC $\approx$0.5 for most of the features, showing their actual uselessness for classification, at least separately from the others. The histograms in Fig. 1 show that the CNN-features were worse on average than the EEG amplitude features. At the same time, there were a relatively small number of CNN-features with a good separating ability.

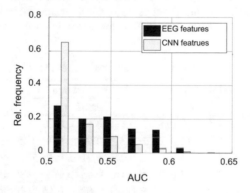

Fig. 1. Distribution of the features by their AUC

Figure 2a shows the dependence of the specificity and sensitivity of the shrinkage linear classifier on the number of the CNN features (in decreasing order of AUC value). Performance measure were averaged over test results obtained with 5-fold cross-validation. The graphs show that the increase in the number of features leads to weak increase of the classifier's sensitivity, which reached a maximum value of 27%.

Thus, the use of only CNN features for classification did not allow us to outperform the results achieved in [2]. However, these features could be used together with the EEG amplitude features.

Figure 2b presents the same dependencies but with the EEG amplitude features used as the classifier's input in addition to the amplitude features. Graphs reveal that the CNN features, taken in number of approximately 30 to 50, increase the sensitivity up to 40% with almost unchanged specificity about 90%. Further increase of the number of added CNN features did not lead to any substantial increase of sensitivity (not shown in the graph).

The Table 1 shows the classification results in the space of $L*T = 169$ amplitude features and 50 CNN-features (total 219 features) in comparison with the results obtained in [2] where the classifier was trained only on the EEG amplitude features (in parentheses). The values are presented in the form $m \pm \sigma$, where m and σ are the mean

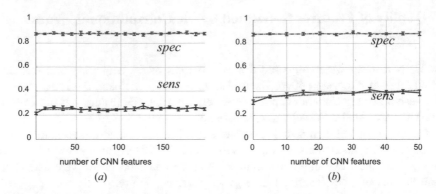

Fig. 2. Specificity and sensitivity of binary classificator as the function of the number of the CNN features: (*a*) only CNN features were used; (*b*) CNN features were used together with EEG amplitude features. For each mean value, a 90%-confidence interval is shown

Table 1. The classification results

Subject	Test on fixations with 500 ms threshold	
	Specificity	Sensitivity
1	0,83 ± 0,18 (0,87 ± 0,13)	0,51 ± 0,20 (0,23 ± 0,17)
2	0,91 ± 0,12 (0,85 ± 0,10)	0,40 ± 0,20 (0,29 ± 0,22)
3	0,83 ± 0,16 (0,84 ± 0,09)	0,51 ± 0,25 (0,51 ± 0,24)
4	0,88 ± 0,08 (0,92 ± 0,12)	0,43 ± 0,20 (0,29 ± 0,21)
5	0,85 ± 0,10 (0,83 ± 0,14)	0,62 ± 0,15 (0,38 ± 0,11)
6	0,91 ± 0,04 (0,87 ± 0,09)	0,35 ± 0,19 (0,23 ± 0,06)
7	0,96 ± 0,04 (0,93 ± 0,05)	0,18 ± 0,09 (0,15 ± 0,09)
8	0,92 ± 0,06 (0,86 ± 0,11)	0,44 ± 0,22 (0,43 ± 0,18)
Mean	0,89 (0,87)	0,43 (0,31)
Std.	0,11 (0,11)	0,22 (0,20)

and standard deviations on test samples, respectively, calculated from the results of fivefold cross-validation. Training and testing of the classifier was carried out for each of the 8 subjects separately. The target EEG signals correspond to the mental intention to click on the screen button using a gaze fixation in the game *EyeLines* [2].

5   Conclusions

The use of CNN-features made it possible to increase the sensitivity of the shrinkage linear classifier of EEG signals used in the EBCI by more than a third, from 31% to 43%, with almost unchanged specificity 90%. The obtained result demonstrates the efficiency of the features formed by the convolutional neural network trained on big data even with respect to the essentially different classification task.

Acknowledgements. This work was supported by Russian Foundation for Basic Research, grant 15-29-01344.

References

1. Protzak, J., Ihme, K., Zander, T.: A passive brain-computer interface for supporting gaze-based human-machine interaction. In: Stephanidis, C., Antona, M. (eds.) Universal Access in Human-Computer Interaction. Design Methods, Tools, and Interaction Techniques for eInclusion, pp. 662–671. Springer, New York (2013)
2. Shishkin, S.L., et al.: EEG negativity in fixations used for gaze-based control: toward converting intentions into actions with an eye-brain-computer interface. Front. Neurosci. **10** (2016)
3. Wolpaw, J.R., Birbaumer, N., McFarland, D.J., Pfurtscheller, G., Vaughan, T.M.: Brain–computer interfaces for communication and control. Clin. Neurophysiol. **113**(6), 767–791 (2002)
4. LeCun, Y., Bengio, Y.: Convolutional networks for images, speech, and time series. Handbook Brain Theory Neural Netw. **3361**(10), 1995 (1995)
5. Krizhevsky, A., Sutskever, I., Hinton, G.E.: Image net classification with deep convolutional neural networks. Adv. Neural. Inf. Process. Syst. **25**, 1097–1105 (2012)
6. Abdel-Hamid, O., et al.: Convolutional neural networks for speech recognition. IEEE/ACM Trans Audio Speech Lang. Process. **22**(10), 1533–1545 (2014)
7. Cecotti, H., Graser, A.: Convolutional neural networks for P300 detection with application to brain-computer interfaces. IEEE Trans. Pattern Anal. Mach. Intell. **33**(3), 433–445 (2011)
8. Subasi, A., Ercelebi, E.: Classification of EEG signals using neural network and logistic regression. Comput. Methods Programs Biomed. **78**(2), 87–99 (2005)
9. Mirowski, P.W., et al.: Comparing SVM and convolutional networks for epileptic seizure prediction from intracranial EEG. In: IEEE Workshop on Machine Learning for Signal Processing, MLSP 2008, pp. 244–249. IEEE (2008)
10. Blankertz, B., Lemm, S., Treder, M., Haufe, S., Müller, K.R.: Single-trial analysis and classification of ERP components – a tutorial. Neuroimage **56**, 814–825 (2011)
11. ImageNet. http://www.image-net.org
12. Russakovsky, O., Deng, J., Su, H., et al.: ImageNet large scale visual recognition challenge. Int. J. Comput. Vis. (IJCV) **115**(3), 211–252 (2015)
13. Wang, Z., Oates, T.: Encoding time series as images for visual inspection and classification using tiled convolutional neural networks. In: Workshops at the Twenty-Ninth AAAI Conference on Artificial Intelligence (2015)
14. Haykin, S.S., et al.: Neural networks and learning machines. Pearson, Upper Saddle River (2009)

Semi-empirical Neural Network Model of Real Thread Sagging

A.N. Vasilyev$^{(\boxtimes)}$, D.A. Tarkhov, V.A. Tereshin, M.S. Berminova, and A.R. Galyautdinova

Peter the Great St. Petersburg Polytechnic University, Saint Petersburg, Russia
a.n.vasilyev@gmail.com, dtarkhov@gmail.com

Abstract. We propose a new approach to building multilayer neural network models of real objects. It is based on the method of constructing approximate layered solutions for ordinary differential equations (ODEs), which has been successfully applied by the authors earlier. The essence of this method lies in the modification of known numerical methods for solving ODEs and their application to an interval of variable length. Classical methods give as a result a table of numbers; our methods provide approximate solutions as functions. This allows refining the model as new information becomes available. In accordance with the proposed concept of building models of complex objects or processes, this method is used by the authors to build a neural network model of a freely sagging real thread. We obtained measurements by conducting experiments with a real hemp rope. Initially, we constructed a rough rope model as a system of ODEs. It turned out that the selection of unknown parameters of this model does not allow capturing the experimental data with acceptable accuracy. Then three approximate functional solutions were built with the use of the authors' method. The selection of the same parameters for two solutions allowed us obtaining the approximations, corresponding to experimental data with accuracy close to the measurement error. Our approach illustrates a new paradigm for mathematical modeling. From our point of view, boundary value problems, experimental data, etc. are considered as raw material for the construction of a mathematical model which accuracy and complexity are adequate to baseline data.

Keywords: Ordinary differential equation (ODE) · Boundary value problem · Approximate solution · Multi-layered solution · Mathematical model · Semi-empirical model · Neural network · Experimental data · Refinement of the model

1 Introduction

It is often impossible when modeling real objects quite accurately describe what happens in physical processes. As a result, an arbitrarily accurate solution of differential equations (which are considered as mathematical models of the mentioned physical processes) does not allow building an adequate mathematical model of the investigated object. In these circumstances, it is reasonable to clarify the object model based on the observation data above it.

© Springer International Publishing AG 2018
B. Kryzhanovsky et al. (eds.), *Advances in Neural Computation, Machine Learning, and Cognitive Research*, Studies in Computational Intelligence 736,
DOI 10.1007/978-3-319-66604-4_21

Refinement both the physics models and the corresponding models in the form of differential equations is a challenging task.

We propose another approach consisting of two stages. In the first step, we construct an approximate solution of the considered differential equations in the form of the function for which the task parameters are the input variables. In the second stage, this function is refined according to observations, the refinement process can continue as new data arrives.

Our approach differs from that of [1–3]; using our semi-empirical model (e.g. neural network model) we replace not a part of the system, which is difficult to simulate by differential equations, but the entire system, including differential equations. This approach is preferred in situations when the accuracy of the description of an object using differential equations is low.

In the paper, this approach is illustrated by the task of calculating sag line of hemp rope (thread) which is hard to solve by standard methods. It is possible to act similarly in many other practically interesting problems.

2 Semi-empirical Model of a Sagging Thread. Methods

We consider hanging freely an inextensible thread of length l, fixed at their ends at the same level. The experiment shows that in the case of strong sagging the conventional model of sagging in a catenary bad describes the experimental data.

In order to determine the line of sagging with account for the bending stiffness, we use the ordinary differential equation

$$\frac{d^2\theta}{ds^2} = \frac{mg}{EJ}(0.5 - s/l)\cos\theta + \frac{A}{EJ}\sin\theta \tag{1}$$

with boundary conditions: $\theta(0) = 0$, $\theta(l) = 0$, and the system of equations

$$\begin{cases} \frac{dx}{ds} = \cos\theta, \\ \frac{dy}{ds} = \sin\theta. \end{cases} \tag{2}$$

Here, E is the Young modulus of the rope material, J is the moment of inertia of the cross-section, L is the distance between supports, s is the length of the portion of the curve, θ is the tangent angle measured from the direction of the horizontal x-axis counterclockwise, A and $B = mg/2$ are the reaction forces of the supports, $q = mg/l$ is the distributed load caused by the weight of the thread, m is the weight of the thread, and g is the magnitude of the gravitational acceleration.

Let us make the change of the variable $t = 2s/l - 1$. In this case the Eq. (1) takes the form

$$\frac{d^2\theta}{dt^2} = -at\cos\theta + b\sin\theta \tag{3}$$

Constants – a and b parameters – are unknown and to be determined from experimental data.

The Eq. (3) is supplemented with the boundary conditions $\theta(0) = \theta(1) = 0$.

The consideration of this issue based on our neural network approach [4–9] is the main content of the article.

The study of the behavior of solutions of the Eq. (3) showed that the desired shape of the thread cannot be obtained.

Further, some options of multi-layered approach for constructing approximate solutions of ordinary differential equations (ODEs) are used [10–15]. The application of the explicit methods to the given task does not lead to the construction of the solution which accurately fits the experimental data. We will apply to the task the implicit Euler method with one step. To do this, we transform the ODE (3) to the system of ODEs

$$\begin{cases} \dfrac{d\theta}{dt} = \phi, \\ \dfrac{d\phi}{dt} = -a\,t\cos\theta + b\sin\theta. \end{cases}$$

The implicit Euler method with one step for this system has the form

$$\begin{cases} \theta = \theta_0 + t\,\phi, \\ \phi = \phi_0 - a\,t^2\cos\theta + b\,t\sin\theta, \end{cases}$$

where $\theta_0 = \theta(0) = 0$, $\phi_0 = \phi(0)$.

The result is $\theta = t\,\phi_0 - a\,t^3\cos\theta + b\,t^2\sin\theta$. Accounting for the boundary condition $\theta(1) = 0$ gives $\phi_0 = a$, where for the angle we obtain the equation

$$\theta = t\,a(1 - t^2)\cos\theta + b\,t^2\sin\theta. \tag{4}$$

The equations for the curve coordinates will be received by integrating (2) using the Simpson method

$$x(s,a,b) = \frac{s}{6m}\left(1 + \cos\theta(s,a,b) + 4\sum_{i=1}^{m}\cos\theta\left(\frac{s}{2m}(2i-1),a,c\right) + 2\sum_{i=1}^{m-1}\cos\theta\left(\frac{s}{m}i,a,b\right)\right);$$

$$y(s,a,b) = \frac{s}{6m}\left(\sin\theta(s,a,b) + 4\sum_{i=1}^{m}\sin\theta\left(\frac{s}{2m}(2i-1),a,c\right) + 2\sum_{i=1}^{m-1}\sin\theta\left(\frac{s}{m}i,a,b\right)\right).$$

We carry out the identification of the parameters a and b by the minimization of the error functional which contains the observation data $\{x_i, y_i\}_{i=1}^{M}$

$$J = \sum_{i=1}^{M} (x(s_i, a, b) - x_i)^2 + \sum_{i=1}^{M} (y(s_i, a, b) - y_i)^2. \tag{5}$$

For this we use an approximate solution of the Eq. (4) constructed in one of three ways.

In the first method we use the approximate equalities $\cos \theta \approx 1$, $\sin \theta \approx \theta$. Thus, from (4) we obtain $\theta_1(t, a, b) = \frac{ta(1-t^2)}{1 - bt^2}$.

In the second method, we use the approximate equalities $\cos \theta \approx 1 - \theta^2/2$, $\sin \theta \approx \theta$. Thus, from the Eq. (4), we obtain $\theta_2(t, a, b) = \frac{2\theta_1(t,a,b)}{1 + \sqrt{1 + \theta_1^2(t,a,b)}}$.

The third method consists of two stages. In the first stage, the Eq. (4) is solved using a neural network. For this purpose the equation is rewritten into the form $\theta = \alpha \cos \theta + \beta \sin \theta$, and the neural network $\theta_3(\alpha, \beta)$ is selected by RProp method of minimizing a sequence of functionals

$$\sum_{i=1}^{m_1} (\theta_3(\alpha_i, \beta_i) - \alpha_i \cos \theta_3(\alpha_i, \beta_i) - \beta_i \sin \theta_3(\alpha_i, \beta_i))^2.$$

Trial points are regenerated as uniformly distributed on the set $[0; 10] \times [0; 1]$ through several steps of the minimizing process (in the experiment, every 5 steps).

We have tested different types of neural networks, but the best result was demonstrated by perceptron with one hidden layer and activation functions $\tanh[a_i\alpha] \tanh[b_i(\beta - d_i)]$.

In the second stage, the parameters a and b are chosen by the minimization of the functional (5).

3 Calculation

Below in Fig. 1 we show (due to symmetry) the experimental points and the right half of the graph for the sag curve of the rope as the resulting application of each of the three methods used.

It should be noted that, despite the general conformity of the curve, the errors in the first method remain too large. In the case of the second method, the shape of the curve, in accordance with the results of the experiment, is significantly better. In the case of the third method, the correspondence of the calculated sag curve of the thread to the results of the experiment was even better. We have got the best result in the second stage of the third method when first we used the neural network with 6 neurons in the form

$$- 0.451 \tanh [0.1\alpha] \tanh [1.4(-1.16 + \beta)] + 0.356 \tanh [1.58\alpha] \tanh[0.413(-0.61 + \beta)]$$
$$+ 0.092 \tanh [3.67\alpha] \tanh [1.28(-0.52 + \beta)] + 0.26 \tanh [7.33\alpha] \tanh [1.013(-0.22 + \beta)]$$
$$- 0.199 \tanh [0.649\alpha] \tanh [2.73(-0.16 + \beta)] + 1.344 \tanh [0.824\alpha] \tanh [1.59(0.948 + \beta)].$$

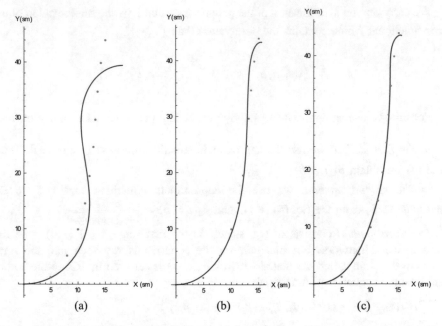

Fig. 1. The right half of the graph of the line $\{x(s, a, b), y(s, a, b)\}$ in the case of (a) the first, (b) the second, (c) the third method and the experimental points.

4 Conclusions

The construction of a mathematical model of an object or process is related, as a rule, to the approximate solution of differential equations with inaccurately specified coefficients for heterogeneous data, which can include experimental data. In a *good* situation (when the equations are chosen well), clarifying the solution of the problem leads to both the reduction of the error of satisfying the equation and the reduction of the error of satisfying the data of measurements. However, if we have not the most successful choice of equations, it is possible to improve the model due to the selection of adjustable parameters to meet the experimental data. This was the case in our model problem. The increase in the number of neurons can improve the accuracy of the solution of the Eq. (4), which, apparently, is not the best way to describe the sagging of the real thread (hemp rope). In this case, the accordance of the solution to the experimental data worsens. This is due both to the approximate compliance of Eq. (1) with the modeled thread and to the inexact transition from the Eq. (1) to (4) because of an insufficient number of layers in the multi-layered approach. The reasonability of increasing the number of layers is questionable due to the inaccuracy of the Eq. (4).

References

1. Prostov, Y.S., Tiumentsev, Y.V.: A hysteresis micro ensemble as a basic element of an adaptive neural net. Opt. Mem. Neural Netw. **24**(2), 116–122 (2015)
2. Egorchev, M.V., Tiumentsev, Y.V.: Learning of semi-empirical neural network model of aircraft three-axis rotational motion. Opt. Mem. Neural Netw. **24**(3), 210–217 (2015)
3. Kozlov, D.S., Tiumentsev, Y.V.: Learning of semi-empirical neural network model of aircraft three-axis rotational motion. Opt. Mem. Neural Netw. **24**(4), 279–287 (2015)
4. Vasilyev, A.N., Tarkhov, D.A.: Mathematical models of complex systems on the basis of artificial neural networks. Nonlinear Phenom. Complex Syst. **17**(3), 327–335 (2014). http://www.j-npcs.org/online/vol2014/v17no3p327.pdf
5. Budkina, E.M., Kuznetsov, E.B., Lazovskaya, T.V., Leonov, S.S., Tarkhov, D.A., Vasilyev, A.N.: Neural network technique in boundary value problems for ordinary differential equations. In: Cheng, L. et al. (eds.) ISNN 2016. LNCS, vol. 9719, pp. 277–283. Springer, Switzerland (2016)
6. Gorbachenko, V.I., Lazovskaya, T.V., Tarkhov, D.A., Vasilyev, A.N., Zhukov, M.V.: Neural network technique in some inverse problems of mathematical physics. In: Cheng, L. et al. (eds.) ISNN 2016. LNCS, vol. 9719, pp. 310–316. Springer, Switzerland (2016)
7. Shemyakina, T.A., Tarkhov, D.A., Vasilyev, A.N.: Neural network technique for processes modeling in porous catalyst and chemical reactor. In: Cheng, L. et al. (eds.) ISNN 2016. LNCS, vol. 9719, pp. 547–554. Springer, Switzerland (2016)
8. Kaverzneva, T., Lazovskaya, T., Tarkhov, D., Vasilyev, A.: Neural network modeling of air pollution in tunnels according to indirect measurements. J. Phys. **772** (2016). http://iopscience.iop.org/article/10.1088/1742-6596/772/1/012035
9. Lazovskaya, T.V., Tarkhov, D.A., Vasilyev, A.N.: Parametric neural network modeling in engineering. Recent Patents Eng. **11**(1), 10–15 (2017)
10. Lazovskaya, T., Tarkhov, D.: Multilayer neural network models based on grid methods. IOP Conf. Ser. Mater. Sci. Eng. **158** (2016). http://iopscience.iop.org/article/10.1088/1757-899X/158/1/01206
11. Bolgov, I., Kaverzneva, T., Kolesova, S., Lazovskaya, T., Stolyarov, O., Tarkhov, D.: Neural network model of rupture conditions for elastic material sample based on measurements at static loading under different strain rates. J. Phys. Conf. Ser. **772** (2016). http://iopscience.iop.org/article/10.1088/1742-6596/772/1/012032
12. Filkin, V., Kaverzneva, T., Lazovskaya, T., Lukinskiy, E., Petrov, A., Stolyarov, O., Tarkhov, D.: Neural network modeling of conditions of destruction of wood plank based on measurements. J. Phys. Conf. Ser. **772** (2016). http://iopscience.iop.org/article/10.1088/1742-6596/772/1/012041
13. Vasilyev, A., Tarkhov, D., Bolgov, I., Kaverzneva, T., Kolesova, S., Lazovskaya, T., Lukinskiy, E., Petrov, A., Filkin, V.: Multilayer neural network models based on experimental data for processes of sample deformation and destruction. In: Selected Papers of the First International Scientific Conference Convergent Cognitive Information Technologies (Convergent 2016), pp. 6–14. Moscow, Russia, 25–26 November (2016). http://ceur-ws.org/Vol-1763/paper01.pdf - Scopus

14. Tarkhov, D., Shershneva, E.: Approximate analytical solutions of mathieu's equations based on classical numerical methods. In: Selected Papers of the XI International Scientific-Practical Conference Modern Information Technologies and IT-Education (SITITO 2016), pp. 356–362. Moscow, Russia, 25–26 November (2016). http://ceur-ws. org/Vol-1761/paper46.pdf - Scopus
15. Vasilyev, A., Tarkhov, D., Shemyakina, T.: Approximate analytical solutions of ordinary differential equations. In: Selected Papers of the XI International Scientific-Practical Conference Modern Information Technologies and IT-Education (SITITO 2016), pp. 393–400 Moscow, Russia, 25–26 November (2016). http://ceur-ws.org/Vol-1761/paper50.pdf - Scopus

Cognitive Sciences and Adaptive Behavior

Color or Luminance Contrast – What Is More Important for Vision?

Evgeny Meilikov[1,2(✉)] and Rimma Farzetdinova[1]

[1] National Research Centre "Kurchatov Institute", 123182 Moscow, Russia
meilikhov@yandex.ru
[2] Moscow Institute of Physics and Technology, 141707 Dolgoprudny, Russia

Abstract. In the framework of a simple analytical model, we *quantitatively* validate the statement that the "color world" is amenable to much more accurate and faster segmentation than the "gray world". That results in significant facilitating conditions required for originating indispensable pop-out effect, and, probably, forms the basis of various cognitive phenomena connected with the color vision. Besides, we show that the known (from optics) Rayleigh criterion for separability of two gray objects is considerably softened for objects of different colors.

Keywords: Pop-out effect · Luminance and color contrast

1 Introduction

Fast analysis of complex images requires high-speed separation of an object from the background that is based on registering important variations of some physical characteristics of the image under jumping from the object to the background (and *vice versa*), that is the local contrast of this parameter while going over the object boundary [1]. Such feature may be the image color, the size or the form of its details, etc.

As a rule, any of those characteristics could be described by some associated physical parameters: for example, the color – by the wavelength (or the set of wavelengths), the form – by the sum of geometrical characteristics, such as the curvature, tilt, symmetry and so on [2]. In any case, the occurrence of image parts, where the value of that parameter deeply differs from its value in the adjacent parts of the image, is the necessary condition for the fast and reliable ("automatic") separating the object [3–5]. It is the essence of the so called "pop-out effect", when the perceptually "evident" region (or the image detail) is readily and spontaneously separated at a glance.

If figure features are not unique, there is no easy separation of the object. Some threshold, characteristic for every feature, is likely exists whose exceedance is required for actuating the pop-out effect. For instance, color or geometrical difference between specific stimulus and distractors needs to be strong enough [3], and if there are diverse differences, the strongest one comes into action [6,7].

© Springer International Publishing AG 2018
B. Kryzhanovsky et al. (eds.), *Advances in Neural Computation, Machine Learning, and Cognitive Research*, Studies in Computational Intelligence 736,
DOI 10.1007/978-3-319-66604-4_22

In some works, attempts have been undertaken to control the efficiency of the visual searching by varying degree of the difference between the stimulus and distractors [8].

The ability to orientate in a complicated exterior situation is of important biological significance, because it provides possibility for a human being and many animals to exist successfully within a fast-changing surrounding. One of challenges in this path is the segmentation of the visual field into the background and the foreground. We do not know in details how the human's (or animal's) brain makes that. Possibly, some mechanisms of the spatial (binocular) vision and various mechanisms of the color vision play considerable role in that operation. In the same time, there are various computer programs that have been developed for solving that problem (see [9], for example) and base on distinguishing boundaries of some image parts [10] (so called the process of the image segmentation). However, for effective action those programs frequently need of some "pointings" from a human.

In the present work, we do not set a goal to investigate psycho-physiological mechanisms of binocular and color vision. The analytic model, which we present, is only aimed to phenomenological (but analytical!) consideration of the above-mentioned pop-out effect in observing two-dimensional images.

To this end, one should define what is the "background". The difficulty is that there are situations, when the background could not be unambiguously determined because the result depends on the context. As a rule, the background keeps less details than the foreground which is just somebody (something) showing up. This "something" differs from the background by relative abundance of details, and, particularly, by abundance of boundaries between regions with distinguishing physical parameters (color, form, etc.). At each point of such a boundary, the absolute value of the gradient $g_i = |\Delta p_i / \Delta n|$ of a given parameter p_i along the normal to the boundary takes much bigger value than in other points.[1] Averaging the gradient value over the region on the order of the mean object size, we discover that the average value $\langle g_i \rangle$ is relatively high in that image part, where we see a lot of details, and is low where the number of details is smallish.

Thus, the pop-out effect consist of following basic stages: (i) foreground, (ii) producing the map for the relevant physical parameter with the smeared (devoid of details) background,[2] and (ii) registering the only locus of activation on that map and perception of the stimulus.

All specified processes are based on the parallel data processing, and provide the principal speed of the pop-out effect.

2   Contrast of a Color Image

Proceed now to quantitative analysing the contrast of color images by the example of the problem concerning distinguishing (fixing the location) the boundary

---

[1] In the above relation, Δn is some finite length along the normal to the boundary, which is small when compared to the object size.

[2] In computer image recognition, that process is termed as pre-processing and is used to reduce a noise by means of its averaging with some filter.

between stripes of *different* colors. It is the extension of Rayleigh resolution limit for two close light spots of *identical* colors. In the framework of our model, we rely from the assumption that luminance and color characteristics do not mix with each other (that is, we do not take into account that, for example, mixing opponent colors could eliminate chromaticity with conserving non-zero luminance, etc.) [11]. These two local, depending on coordinates, image features could be associated with two separate, mutually not depending, parameters (kind of the amplitude and the phase of the wave function in quantum mechanics). Let the image be planar and two-colored (that is, the picture radiates or reflects the light of two, subjectively strongly differing, wavelengths) and let that image presents a set of smeared, parallel to each other, stripes of two different colors (red and yellow – to be definite). For the sequel, it is very convenient to consider a given image to be consisting of a number red (r) and yellow (y) points whose densities arise the subjective color sensation.

For two colors considered, one could introduce two local complex parameters $A_r(\mathbf{r})e^{i\varphi_r}$ and $A_y(\mathbf{r})e^{i\varphi_y}$ depending of coordinates $\mathbf{r}$. Here, $A_r(\mathbf{r})$, $A_y(\mathbf{r})$ are spatial dependencies of mean amplitudes of the waves' field intensities for each of two light waves (defined, for instance, by densities of color points within the bitmap image), and φ_r, φ_y are color phases of those two waves (i is the imaginary unit).[3] Those phases are, in fact, "codes" of mixed colors – conditional parameters that, in the framework of the simple suggested model, possess following properties: (1) they do not depend on light intensities of chosen colors but depend on their wavelengths only, (2) complex color parameters of each small region of two-colored raster image are additive (generalized Abny law of luminance additivity) [11], and the resulting complex parameter $A(\mathbf{r})e^{i\varphi(\mathbf{r})}$, including the summary color amplitude (luminance) $A(\mathbf{r})$ and the summary color phase $e^{i\varphi(\mathbf{r})}$, is determined according to the relationship (being based on rules of complex numbers' or vectors' addition)

$$A(\mathbf{r})e^{i\varphi(\mathbf{r})} = A_r(\mathbf{r})e^{i\varphi_r} + A_y(\mathbf{r})e^{i\varphi_y}, \tag{1}$$

where φ_r, φ_y are phases of red and yellow colors. That leads to

$$A(\mathbf{r}) = \sqrt{A_r^2(\mathbf{r}) + A_y^2(\mathbf{r}) + 2A_r(\mathbf{r})A_y(\mathbf{r})\cos\Delta\varphi}, \quad \varphi(\mathbf{r}) = \varphi_r + \arcsin\left[\frac{A_y(\mathbf{r})}{A(\mathbf{r})}\sin\Delta\varphi\right], \tag{2}$$

where $\Delta\varphi = \varphi_y - \varphi_r$. As one could expect, $\varphi = \varphi_r$ at $A_y = 0$ and $\varphi = \varphi_r + \Delta\varphi = \varphi_y$ at $A_r = 0$. In the general case, (A_r, $A_y \neq 0$) one has $\varphi_r < \varphi < \varphi_y$, that corresponds to the second Grassmann law – mixing two colors results in the color lying (on the color ring) between mixed colors.

---

[3] We restrict ourselves to the case of two-color images, to which the plain (two-dimensional) space of color tones corresponds. Such a space could be described by the combination of two complex numbers. Within that space, any color, represented by the sum of yellow and red colors of varied intensities, is displayed as the point on the color plane with zero contribution of the blue color. Considering three-color (and, hence, three-dimensional) space would result in the significant complicating the model.

In the theory of the color vision, the polar angle of the color ring (measured in radians or degrees) is the analog of the color phase φ, so that in the course of numerical calculations one should proceed from the consistency between the color phase and the light wave length λ. Though that consistency is subjective, the required empiric correspondence $\varphi(\lambda)$ is known [11]. That relation based on the traditional view of the color ring, where the angle interval from 0 to 360° corresponds to te visible spectrum. According to that scheme, the color phase *rises* monotonously with transition from the blue color to the red one (that is, with increasing the wave length), and the red color is supposed to relate with the polar angle $\varphi_r = 2\pi$. However, it is clear that the specific phase value φ_r could be chosen arbitrary, basing on considerations of convenience or tradition only. For instance, in physical optics the wave phase is proportional not to the wavelength, but to some conversed value, the so called wave number $k \propto 1/\lambda$, and so it *falls down* with passing from the blue color to the red one. Thus, below we accept $\varphi_r = 0$, and the empirical dependence $\varphi(\lambda)$ [11] has been reconstructed using the simple linear substitution $\varphi \to 2\pi - \varphi$.

Such a reconstructed dependence $\varphi(\lambda)$ is shown in Fig. 1. It has the two-step form with parts of the most pronounced change being centered near wave lengths of $\lambda_1 \approx 485$ (blue color) and $\lambda_2 \approx 580$ nm (yellow color), and each of those sections have the width $2\,\delta\lambda = 20$ nm. That empirical dependence is well approximated by the sum of two sigmoid functions:

$$\varphi(\lambda) = \left(\pi - \frac{1}{2}\right)\left\{2 - \frac{1}{1 + \exp\left[-(\lambda - \lambda_1)/\delta\lambda\right]} - \frac{1}{1 + \exp\left[-(\lambda - \lambda_2)/\delta\lambda\right]}\right\} \quad (3)$$

Three plateaus of that dependence correspond to basic colors of the RGB-model (red, green, and blue). Within those plateaus, color sensations are stable – they do not depend (over a wide range) on the wave length, and are based on the properties of corresponding visual sensors (cone cells, reacting on these three colors.) Why those plateaus are needed? Possible answer is related the volatility of natural illumination which depends on the sun angle and the atmosphere

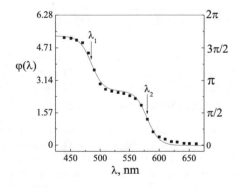

Fig. 1. Empirical function $\varphi(\lambda)$. Points are experimental data [17], solid line is the fitting function (3).

state (clear, rain, fog). With cloudy weather, the light is more cold – the color temperature shifts to the blue (short-wave length) spectrum part. With bright sun, colors are more warm – it contains more red and yellow rays. Three plateaus provide stability of visual sensations: within the broad range of conditions, red berries stay red, green grass stays green and so on, that assists in foraging.

According to Fig. 1 (and Eq. (3)), the color phase of the red color equals $\varphi_r = 0$, for the yellow light $\varphi_y \approx \pi/2$, for the green light $\varphi_g \approx \pi$, and for the violet one $\varphi_b \approx 2\pi$. That coding style is, really, neither more nor less than the analytic version of the known color space DKL [12,13], where the color is coded by one of the color space angle (changing from 0 to 2π).

For simplicity, we consider below the image of two colors – red and yellow, for which $\varphi_r = 0$, $\Delta\varphi = \varphi_y - \varphi_r = \pi/2$. In that case, Eq. (2) takes the simpler form[4]:

$$A(\mathbf{r}) = \sqrt{A_r^2(\mathbf{r}) + A_y^2(\mathbf{r})}, \quad \varphi(\mathbf{r}) = \arcsin\left[\frac{A_y(\mathbf{r})}{A_r(\mathbf{r}) + A_y(\mathbf{r})}\right]. \tag{4}$$

The first of these relations corresponds to the Abby law of luminance additivity [11], and the second one, as required, leads to $\varphi = 0$ at $A_y = 0$ (the red color only) and $\varphi = \pi/2$ at $A_r = 0$ (the yellow color only).

Thus, two spatial functions are associated with color picture: the color amplitude $A(\mathbf{r})$ and the color phase $\varphi(\mathbf{r})$. Subjective perception of a picture is attained by registering variations of those functions and, mostly, that one whose relative variations are bigger.

In this regard, let us discuss two examples. The first one is the periodical structure of alternating (along x-axis) parallel red and yellow stripes with unit distance between their centers. Each stripe is smeared due to decaying its local luminance out of the center by the Gauss law (centers' coordinates are $x_n = 2n$, $n = 0, \pm 1, \pm 2, \ldots$ – red stripes, $x_n = 2n + 1$, $n = 0, \pm 1, \pm 2, \ldots$ – yellow stripes):

$$A_r(x - x_n), \ A_y(x - x_n) = F_n(x), \quad F_n(x) = A_0 \exp\left[-\frac{(x - x_n)^2}{2\sigma^2}\right], \tag{5}$$

where A_0 is the light intensity in stripes' centers, and σ is the effective width of smeared color stripes. According to Eq. (4) color amplitude and phase in any point x of the line, perpendicular to stripes, equal

$$A(x) = \sqrt{\hat{A}_r^2(x) + \hat{A}_y^2(x)}, \quad \varphi(x) = \arcsin\left[\frac{\hat{A}_y(x)}{\hat{A}_r(x) + \hat{A}_y(x)}\right], \tag{6}$$

---

[4] Simpler forms of Eq. (4) comparing to original relations (2) is the result of the above-made "involuntary" choice of color phases leading to $\varphi_y \approx \pi/2$. One could verify that all conclusions about the contrast of color images remain valid with some another choice of the dependency $\varphi(\lambda)$. This is due to the fact that actual are not numerical values of colors phases, but gradient of the phase along the direction of its steepest variation which is the relative characteristics, analogical to the image contrast (to the map of phase variation, in the case considered).

where

$$\hat{A}_r(x) = A_0 \sum_{n=-\infty}^{\infty} F_{2n}(x), \quad \hat{A}_y(x) = A_0 \sum_{n=-\infty}^{\infty} F_{2n+1}(x), \tag{7}$$

Which of two functions $A(x)$, $\varphi(x)$ varies stronger? Answer to this question depends on the degree of smearing the stripes which is defined by the σ-parameter value. Corresponding dependencies $A(x)$ and $\varphi(x)$ are presented in Fig. 2 for different values $\sigma = 0.2, 0.5, 0.75, 1$. If stripes have sharp edges ($\sigma \ll 1$), then $A(x)$ varies in the range from 0 up to 1 (see Eq. (5)), and $\varphi(x)$ – from 0 to $\pi/2$ (see Eq. (6)), that is, in fact, near equally. Under essential but moderate stripe smearing ($\sigma \lesssim 1$), the relative variation of $A(x)$ amounts a few percents merely, while the phase $\varphi(x)$ continues to vary by two-three times. And at last, with the very strong stripe smearing ($\sigma \gtrsim 1$), when they are significantly overlayed, color parameters do not practically depend on coordinates and the picture represents the faint, nearly uniform, background where one could not discriminate some details.

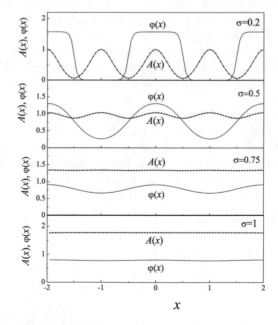

Fig. 2. Spatial dependencies of light intensity $A(x)$ and color phase $\varphi(x)$ for a system of alternating by color (red and green) smeared parallel stripes with various smearing parameters σ (the width of the Gauss smearing).

The quantitative characteristics of the picture visibility is the coefficient of modulation (or the contrast) of some or other parameter defined by the relations $V_A = (A_{max} - A_{min})/(A_{max} + A_{min})$ or $V_\varphi = (\varphi_{max} - \varphi_{min})/(\varphi_{max} + \varphi_{min})$, where indexes min and max are assigned to the minimum and maximum values of a given parameter (in our case – to the light intensity A and color phase φ). As could be seen from Fig. 3, with the moderate smearing of stripes ($\sigma = 0.3 - 0.7$), the contrast of the color phase is by an order of magnitude higher than the intensity contrast. Does it mean that the image visibility and the pop-out effect are associated namely with the color (not luminous) contrast? Generally – not, if corresponding thresholds of perception are also differ strongly (if, for instance, the threshold contrast of luminance is on the order of value lower than that for the color contrast, then differences of coefficients V_A and V_φ are cancelled and the intensity reception becomes to be no less important than the color one). The answer to the question could be provided by experiments.

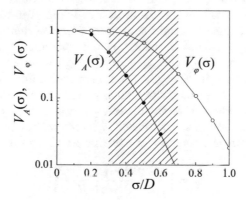

Fig. 3. Luminance V_A and color V_φ contrasts for a system of alternating by color (red and green) smeared parallel stripes with various smearing parameters σ. Shadowed is the diapason of σ/D-values, where the color contrast is higher on the order of value than that of luminance. (D is the distance between stripe centers, σ is the width of the Gauss smearing.)

It is known that human eye is able to differentiate luminance gradations (to have the contrast sensibility) of about 1% [14], that corresponds to the threshold coefficient of the luminance modulation $\delta V_A/V_A \sim 10^{-2}$. Analogically, eye discriminates colors, corresponding to the difference in radiation wavelengths of 1–5 nm [15] (spectral sensibility). That corresponds to the threshold coefficient of the color modulation $\delta V_\varphi/V_\varphi \sim (0.1 - 0.5) \cdot 10^{-2} \lesssim \delta V_A/V_A$.

Specified threshold sensitivities for varying luminance or color relate to ideal conditions of observation. Real thresholds are higher by the order of value but remain to be alike: $\delta V_A/V_A \approx \delta V_\varphi/V_\varphi \sim 0.1$. That allows to conclude that with moderate stripe smearing the picture is percepted more reliably (and as a result – faster) via variations of the color phase (that is, by color variations) and not through the light intensity variations. For example, separating green stripes on the red background is realized more reliably and faster than separating bright-red stripes on the light-red background. That conclusion is validated by experimentally determined functions of the contrast sensitivity which relate to registering modulated by color (red/green or blue/yellow) lattices of the constant luminance [16,17]. Experiments show that for spatial frequencies lower than ~0.2 period/grad (till physiological limitations are not important) the contrast sensibility for observing color (chromatic) lattice excels (up to five times with the spatial frequency of about 0.07 period/grad) the one for the luminance (intensity) lattice. That result is in full agreement with the above-cited conclusion about the higher contrast of chromatic image.

3   Color Analog of Rayleigh Criterion

The next example is the color analog of the Rayleigh criterion [18], which in its classic "gray" variant defines the resolution of human eye for observing a pair of spatially close and smeared light sources of *identical luminance and color*. In that case, according to Rayleigh, the sources are discriminated as separate ones, if the luminance contrast $V_A = [A_{max} - A(0)]/A_{max}$ is not lower than ~0.2 (here $A(0)$, A_{max} are the light intensity in the image center and maximum intensity, correspondingly). For smeared stripes, such a luminance contrast appears under condition $\sigma \lesssim 0.7D$, where D is the distance between stripe centers.

In the "color" variant (two stripes of *different color*, smeared by Gauss), it is convenient to define the contrast of color phase V_φ by the relation

$$V_\varphi = \left| \frac{\varphi(D/2) - \varphi(-D/2)}{\varphi(D/2) + \varphi(-D/2)} \right|, \tag{8}$$

where $\varphi(\pm D/2)$ is the color phase in centres of both (smeared) stripes. In that case, as could be seen in Fig. 4, with decreasing the distance between stripes the color contrast of the boundary drops much slower than the luminance contrast. Fixing the color change on the smeared boundary between stripes is possible and

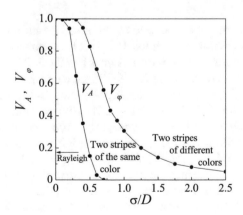

Fig. 4. Luminance V_A and color V_φ contrasts for two smeared parallel stripes of red and green color as a function of the ratio σ/D. (D is the distance between stripe centers, σ is the width of the Gauss smearing.) Arrow corresponds to the Rayleigh criterion.

reliable even in the case when the luminance contrast is not enough to detect the boundary.

That conclusion corresponds to the statement that "color world" is amenable to faster and more accurate segmentation than "gray world". Therein, even those boundaries are resolved which could not be distinguished in gray world according to Rayleigh criterion.

4 Conclusion

In this work, the attempt has been done to ground quantitatively (in the framework of a simple analytic model) the observation that "color world" is amenable to faster and more accurate segmentation than "gray world". That results in significant facilitating conditions required for originating indispensable pop-out effect, and, probably, forms the basis of various cognitive phenomena connected with the color vision.

Besides, we show that the known (from optics) Rayleigh criterion for separability of two gray objects is considerably softened for objects of different colors.

References

1. Wolfe, J.M.: Guided search 2.0: a revised model of visual search. Psychon. Bull. Rev. **1**, 202–238 (1994)
2. Nakayama, K., Silverman, G.H.: Serial and parallel processing of visual feature conjunctions. Nature **320**, 264–265 (1986)
3. Duncan, J., Humphreys, G.W.: Visual search and stimulus similarity. Psychol. Rev. **96**, 433–458 (1989)
4. Bergen, J.R., Julesz, B.: Parallel versus serial processing in rapid pattern discrimination. Nature **303**, 696–698 (1983)
5. Treisman, A.M., Gelade, G.: A feature integration theory of attention. Cogn. Psychol. **12**, 97–136 (1980)
6. Quinlan, P.T.: Visual feature integration theory: past, present, and future. Psychol. Bull. **129**, 643–673 (2003)
7. Search, T.A.: Similarity, and integration of features between and within dimensions. J. Exp. Psychol. Hum. Percept. Perform. **17**, 652–676 (1991)
8. Avraham, T., Yeshurun, Y., Lindenbaum, M.: Predicting visual search performance by quantifying stimuli similarities. J. Vis. **8**, 122 (2008)
9. Hatfield, L.T., Douglas, S.H., Rohring, N.W., Kotler, M., Fikree, M.: Patent US20100158379 A1 (2010)
10. Gonzalez, R., Woods, R.: Digital Image Processing. Prentice-Hall, Inc., Upper Saddle River (2002)
11. Norton, T.T., Corliss, D.A., Bailey, J.E.: Psychophysical Measurement of Visual Function. Richmond Products, Butterworth-Heinemann (2002)
12. Derrington, A.M., Krauskopf, J., Lennie, P.: Chromatic mechanisms in lateral geniculate nucleus of macaque. J. Physiol. **357**, 241–265 (1984)
13. D'Zmura, M., Knoblauch, K.: Spectral bandwidths for the etection of color. Vis. Res. **38**, 3117–3128 (1998)

14. Blackwell, H.R.: Contrast thresholds of the human eye. J. Opt. Soc. Am. **36**, 624–643 (1946)
15. Graham, C.H.: Color: data and theories. In: Graham, C.H., et. al. (eds.) Vision and Visual Perception, N.Y., pp. 414–451 (1965)
16. Rovamo, J.M., Kankaanpaa, M.I., Kukkonen, H.: Modelling spatial contrast sensitivity functions for chromatic and luminance-modulated gratings. Vision. Res. **39**, 2387–2398 (1999)
17. Mullen, K.T.: The contrast sensitivity of human colour vision to red green and blue yellow chromatic gratings. J. Physiol. **359**, 381–400 (1985)
18. Born, M., Wolf, E.: Principles of Optics. Cambridge University Press, Cambridge (1999)

Synchrony of Cortical Alpha and Beta Oscillations

Victor L. Vvedensky[✉]

National Research Center "Kurchatov Institute", Moscow, Russia
victorlvo@yandex.ru

Abstract. We analyzed behavior of cortical alpha and beta oscillations during preparation of self-paced finger movement. Magnetic field patterns around the head were complex during rhythmic events, implying several simultaneously active sources. Oscillations in the cortical sites separated by a few centimeters were highly synchronized. The instants of the maximum amplitude for sharp peaks generated in different locations match each other with millisecond precision. During long trains of oscillations field patterns changed with each new cycle, which means that each time a new combination of cortical sites was activated. We conjecture that the trains of alpha and beta oscillations maintain proper level of synchronization between distant neural populations in the brain. They organize a pool of neurons needed to support normal performance of the cortex executing the task.

Keywords: MEG · Self-initiated movement · Single-trial response · Alpha and beta oscillations · Multiple sources

1 Introduction

The historical view of (EEG/MEG)-measured alpha rhythms (~10 Hz) as a "resting" or "idle" brain state is being challenged by evidence that they are actively and topographically deployed to gate information processing. They can be beneficial to attentional regulation across neocortical areas [1–3]. A number of studies point at the active role of alpha oscillations in the process of selective attention which filters out distractive information [4, 5]. Beta oscillations (~20 Hz) are believed to take part in sensory-motor transmission and to be related to anticipation of an impending event [6]. Alpha and beta rhythms are usually considered separately, since under many experimental conditions they are recorded in different cortical areas. Alpha-rhythm is usually dominant in the visual occipital cortex, beta-rhythm in the sensory-motor parietal areas. Alpha-rhythm in the occipital areas has maximum amplitude when the eyes are closed and becomes blocked when the subject opens the eyes. MEG measurements of spontaneous activity during motor tasks are usually made with eyes open and both alpha and beta components are present in the same somatosensory area [7, 8]. Our experiments show that during the preparation of a voluntary movement alpha and beta oscillations are densely interspersed in the same cortical sites. In our view they support selective attention state preceding arrival of internal stimuli.

© Springer International Publishing AG 2018
B. Kryzhanovsky et al. (eds.), *Advances in Neural Computation, Machine Learning, and Cognitive Research*, Studies in Computational Intelligence 736,
DOI 10.1007/978-3-319-66604-4_23

2   Materials and Methods

Eight right-handed volunteers 22–31 years old took part in the study. None of the participants had known neurological or psychiatric disorders. The study was approved by the local ethics committee of the Moscow University of Psychology and Education and was conducted following the ethical principles regarding human experimentation (Helsinki Declaration). The subjects were instructed to make quick index finger extensions at their own will, keeping the finger in the up position for some time and then moving it back into the original position. Magnetic brain responses were recorded using a helmet-shaped whole-head MEG system (Elekta Neuromag, 306 channels) at the Moscow University of Psychology and Education. During the measurement, the participants were sitting in a lighted magnetically shielded room with eyes open. The MEG signals were recorded with a sampling rate of 1000 Hz and native hardware filters with bandpass 0.1–300 Hz. The main stages of our data acquisition analysis procedures are described in the publications [9–11].

3   Results

We measured magnetic field over the whole head when subjects were performing repetitive self-paced finger movement. During the session, the subjects were always ready to make self-paced finger movement either up or down. They were instructed to avoid automatic movements and to choose waiting time before the movement in each trial separately. Under this condition, rhythmic events in both alpha and beta frequency ranges were consistently observed. Scrutiny of individual records shows that the alpha and beta rhythmic events do not overlap, but follow each other, often being concatenated [11]. MEG measurements under somewhat different conditions, reported in [12], showed that mu-alpha or mu-beta events were nonoverlapping for roughly 50% of their respective durations in single trials. We looked for common features in the observed alpha and beta processes by comparing maps of the measured magnetic field, calculated using Brainstorm software [13]. Field maps for alpha and beta events (trains of oscillations) usually displayed complex structure, often difficult to interpret. This is due to simultaneous activity of many cortical sources, most of which are traditionally considered as noise. For high amplitude events the patterns often displayed regular structure, examples are shown in Fig. 1. They are quite different from the dipolar patterns routinely used in MEG studies for localization of the active area in the brain. The field distributions shown in Fig. 1 are not exceptions - similar combinations occur permanently. Dipolar patterns are rare. Field patterns shown in Fig. 1 and many similar ones imply a small number of active sources in the brain, which cooperate to produce these distributions of the magnetic field. We record our signals in the wide frequency band 0.1–300 Hz, so the shape of the signals is not distorted. Traces often look like triangles with sharp corner reiterating with alpha or beta frequency. Sometimes just single triangular peaks are observed.

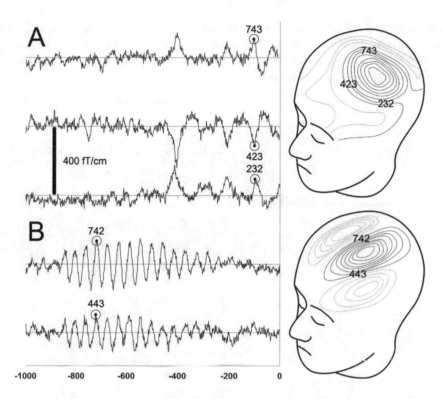

Fig. 1. Single trial magnetic signals in different sensors during a second before right index finger movement. The spatial patterns of the magnetic field over the head are shown for the instants of maximum amplitude of cortical oscillations (marked with *open circles*). Amplitude of the radial component of the magnetic field is represented as isomagnetic chart with contour step 100 fT. Numbers indicate positions of the sensors over the head. Bandwidth 0.1–300 Hz. A - Peak in the alpha train of oscillations, preceding the movement for 100 ms. Subject 1. B - Peak in the beta train, preceding the movement for 720 ms. Subject 2.

The map at the instant, when the field value reaches the top of the peak, reveals extremely sharp synchronization of several neural populations in the cortex located few centimeters apart. In these regions the growth of electric polarization in large groups of neurons reaches the top value simultaneously with millisecond precision. At this point polarization build-up in several cortical pools of neurons abruptly turns into a fall. This instant is the most appropriate for the generation of neuronal spikes travelling coherently into other cortical areas [14]. A certain group of neurons can receive simultaneous inputs from clearly distinct neural populations, which can recruit this group into a common pool. Recordings of local potentials, using electrodes implanted into basal ganglia in alert patients during functional neurosurgery for Parkinson's disease, show that behavioral events can be reflected in momentary changes in the degree of synchronization between neuronal elements [15]. This momentary synchronization is different from the synchronization of rhythmic events extended in time for several oscillations. Wavelet analysis is the standard tool for the MEG study of the synchrony in the cortex. This

method implies similar behavior of the cortical ensembles of neurons during each oscillation through the whole width of the wavelet, typically 4–7 periods long. In contrast,
we analyzed each single oscillation and see that field patterns vary substantially from
cycle to cycle, as shown in Fig. 2.

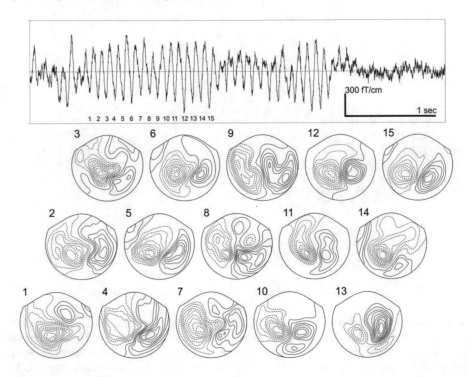

Fig. 2. Single-channel MEG signal over right parieto-occipital part of the head during time
interval before right index finger moved up in the end of the record. This is one of many rhythmic
events with large amplitude during the session. Isomagnetic charts for 15 subsequent peaks of
oscillations in the alpha frequency range are plotted. The maps correspond to the instants of
maximum amplitude. *Top* view on the flattened helmet is shown, nose up. Considerable variations
of the pattern complexity and the changes of the direction of maximum gradient in every cycle
are evident.

Orientation of the cortical source, providing maximum contribution into the field
pattern, changes from cycle to cycle and contributions from other sources appear and
disappear in different cycles in a seemingly irregular way. Averaging across the whole
long spindle of oscillations would produce rather neat picture with a dipolar source in
the right parieto-occipital area, in spite of variations in different cycles. The cortical
process during alpha spindle looks like a sequence of fireworks volleys, where each
cycle is an individual pattern with few active spots in different cortical sites. High
amplitude spindle of alpha peaks is a regular chain of individual events. The area of the
cortex active during each peak can be quite small, as show studies with cortical electrode
arrays 1 cm large [16]. Larger intracranial electrode arrays, covering several centimeters

of cortical surface, reveal multiple cortical spikes scattered over the cortex of patients with excessively synchronized epileptic activity [17]. The trains of alpha and beta oscillations in healthy subjects can maintain proper level of synchronization between distant neural populations in the brain needed to support normal performance of the cortex executing the task.

4 Conclusions

Alpha and beta rhythmic events in the brain, observed when the subject is ready to perform intended action display sharp synchronization between several cortical sites few centimeters apart.

Acknowledgments. The author expresses his gratitude to A.O. Prokofyev, A.Yu. Nikolayeva, A.V. Butorina and T.A. Stroganova who provided their reliable experimental data. Supported by the Grant of the Russian Fund for Basic Research 15-29-03814-ofi_m.

References

1. Jones, S.R., Kerr, C.E., Wan, Q., et al.: Cued spatial attention drives functionally relevant modulation of the mu rhythm in primary somatosensory cortex. J. Neurosci. **30**, 13760–13765 (2010)
2. Dockstader, C., Cheyne, D., Tannock, R.: Cortical dynamics of selective attention to somatosensory events. Neuroimage **49**(2), 1777–1785 (2010)
3. Sacchet, M.D., LaPlante, R.A., Wan, Q., et al.: Attention drives synchronization of alpha and beta rhythms between right inferior frontal and primary sensory neocortex. J. Neurosci. **35**(5), 2074–2082 (2015)
4. Foxe, J.J., Snyder, A.C.: The role of alpha-band brain oscillations as a sensory suppression mechanism during selective attention. Front. Psychol. **2**(154), 1–13 (2011)
5. Palva, S., Palva, J.M.: New vistas for alpha-frequency band oscillations. Trends Neurosci. **30**(4), 150–158 (2007)
6. Kilavik, B.E., Zaepffel, M., Brovelli, A., MacKay, W.A., Riehle, A.: The ups and downs of beta oscillations in sensorimotor cortex. Exp. Neurol. **245**, 15–26 (2013)
7. Gaetz, W., Cheyne, D.: Localization of sensorimotor cortical rhythms induced by tactile stimulation using spatially filtered MEG. Neuroimage **30**, 899–908 (2006)
8. Szurhaj, W., Derambure, P., Labyt, E., et al.: Basic mechanisms of central rhythms reactivity to preparation and execution of a voluntary movement: a stereoelectroencephalographic study. Clin. Neurophysiol. **114**(1), 107–119 (2003)
9. Chaianov, N.V., Prokof'ev, A.O., Morozov, A.A., Stroganova, T.A.: Localization of cortical motor areas in humans by magnetoencephalography. Zh. Vyssh. Nerv. Deiat. **62**(5), 629–640 (2012). Article in Russian
10. Vvedensky, V.L.: Individual trial-to-trial variability of different components of neuromagnetic signals associated with self-paced finger movements. Neurosci. Lett. **569**, 94–98 (2014)
11. Vvedensky, V.L., Prokofyev, A.O.: Timing of cortical events preceding voluntary movement. Neural Comput. **28**, 286–304 (2016)

12. Jones, S.R., Pritchett, D.L., Sikora, M.A., et al.: Quantitative analysis and biophysically realistic neural modeling of the MEG mu rhythm: rhythmogenesis and modulation of sensory-evoked responses. J. Neurophysiol. **102**, 3554–3572 (2009)
13. Tadel, F., Baillet, S., Mosher, J.C., Pantazis, D., Leahy, R.M.: Brainstorm: a user-friendly application for MEG/EEG analysis. Comput. Intell. Neurosci. **2011**, 1–13 (2011). Article ID 879716. http://neuroimage.usc.edu/brainstorm/
14. Steriade, M.: Synchronized activities of coupled oscillators in the cerebral cortex and thalamus at different levels of vigilance. Cereb. Cortex **7**(6), 583–604 (1997)
15. Cassidy, M., Mazzone, P., Oliviero, A., et al.: Movement-related changes in synchronization in the human basal ganglia. Brain **125**(6), 1235–1246 (2002)
16. Vanleer, A.C., Blanco, J.A., Wagenaar, J.B., Viventi, J., Contreras, D., Litt, B.: Millimeter-scale epileptiform spike propagation patterns and their relationship to seizures. J. Neural Eng. **13**(2), 026015 (2016)
17. Tao, J.X., Ray, A., Hawes-Ebersole, S., Ebersole, J.S.: Intracranial EEG substrates of scalp EEG interictal spikes. Epilepsia **46**(5), 669–676 (2005)

Processes of Self-organization in the Community of Investors and Producers

Vladimir G. Red'ko[1,2] and Zarema B. Sokhova[1(✉)]

[1] Scientific Research Institute for System Analysis, Russian Academy of Sciences, Moscow, Russia
vgredko@gmail.com, zarema_s@mail.ru
[2] National Research Nuclear University MEPhI (Moscow Engineering Physics Institute), Moscow, Russia

Abstract. The paper analyzes the processes of self-organization in the economic system that consists of investors and producers. There is intensive information exchange between investors and producers in the considered community. The model that describes the economic processes has been developed. The model proposes a specific mechanism of distribution of investors capital between producers. The model considers the interaction mechanism between investors and producers in a decentralized economic system. The main element of the interaction is the iterative process. In this process, each investor takes into account the contributions of other investors into producers. The model is investigated by means of the computer simulation, which demonstrates the effectiveness of the considered mechanism.

Keywords: Investors · Producers · Decentralized system · Competition · Self-organization · Collective behavior

1 Introduction

Competition is an important element of the economic systems. Is cooperation possible in competitive societies? Based on game theory and computer simulation, Robert Axelrod demonstrated the advantages of cooperation for two players [1]. Forms of aggressive and constructive competition between individuals within an agent-oriented approach were also analyzed in [2]. In the current paper, we design and investigate the model of the economic system with a soft constructive competition. The prototype of our model is the works of Belgian researchers [3, 4]; their systems have used agents-messengers to optimize a production hall's operation and routing car traffic in a city.

In our model, the economic system is the community of producers and investors. The producers and investors compete with each others. Nevertheless, the information about capitals, profits, and intentions of community members is open within the community. In particular, investors inform producers about their intention to invest the certain values of capital into the separate producers. The information exchange ensures the possibility to create a decentralized system of interaction within the community of

© Springer International Publishing AG 2018
B. Kryzhanovsky et al. (eds.), *Advances in Neural Computation, Machine Learning, and Cognitive Research*, Studies in Computational Intelligence 736,
DOI 10.1007/978-3-319-66604-4_24

investors and producers. The iterative process is an important element of the model. This iterative process helps each investor to take into account the intentions of other investors. The model describes an effective interaction of investors and producers in the economic community. This effective interaction was demonstrated by means of computer simulation.

2   Description of the Model

2.1   General Scheme of the Model

We consider a community of N investors and M producers; each of them has a certain capital K_{inv} and K_{pro}. The investors and producers operate in the transparent economic system, i.e. they provide the information about their current capital and profit to the entire community. There are periods of operation of the community. For example, a period can be equal to one year. Further, T is a time period number.

At the beginning of each T period, a particular investor makes an investment into m producers. At the end of the period, every investor has to decide: how much capital should be invested into one or another producer in the next period. In order to take into account the intentions of all investors, we introduce an iterative process, which is described below.

The i-th producer has its own initial capital C_{i0} before the period T. The producer obtains some additional capital from investors. The whole capital of the producer i is:

$$C_i = C_{i0} + \sum_{j=1}^{N} C_{ij}, \tag{1}$$

where C_{ij} is the capital invested into the i-th producer by the j-th investor at the beginning of the period T.

We believe that the dependence of the producer profit R_i on its current capital C_i has the form:

$$R_i(C_i) = k_i F(C_i), \tag{2}$$

where the coefficient k_i characterizes the efficiency of the i-th producer. The values k_i vary randomly at the end of each period. The function $F(x)$ is the same for all producers. In the current work, we believe that the function $F(x)$ has the form:

$$F(x) = \begin{cases} ax, & if\ ax \le Th \\ Th, & if\ ax > Th \end{cases}, \tag{3}$$

where Th is the threshold of the function $F(x)$.

At the end of the period T, the producer returns the invested capital to its investors. In addition, the producer pays off a part of its profit to the investors. The j-th investor receives the profit part that is proportional to the investment made into this producer:

$$R_{ij} = k_{repay}R_i(C_i)\frac{C_{ij}}{\sum_{l=1}^{N} C_{il}}, \tag{4}$$

where C_i is the current capital of the i-th producer, k_{repay} is the parameter determining the part of the profit that is transferred to investors, $0 < k_{repay} < 1$. The producer itself gets the remaining part of the profit:

$$R_i^* = (1 - k_{repay})R_i(C_i). \tag{5}$$

Each investor has the following agents-messengers: the searching agents and the intention agents; these agents are used for information exchange within the community.

2.2 Description of the Iterative Process

At the first iteration, the investor sends the searching agents to all producers in order to determine the current capital of each producer. At the first iteration, the investor does not take into account the intentions of other investors to invest some capitals into producers. The investors estimate the values A_{ij}, which characterize the profit expected from the i-th producer in the next period T. These values A_{ij} are:

$$A_{ij} = k_{dist}R_{ij} = k_{dist}k_{repay}k_iF(C'_{i0})\frac{C_{ij}}{\sum_{l=1}^{N} C_{il}}, \tag{6}$$

where C_{il} is the capital invested into the i-th producer by the l-th investor, C'_{i0} is the expected initial capital of the i-th producer at the beginning of the next period, $k_{dist} = k_{tested}$ or $k_{untested}$ ($k_{tested} > k_{untested}$). The positive parameters k_{tested}, $k_{untested}$ indicate the level of the confidence of the investor for the considered producer; this level of confidence is k_{tested} and $k_{untested}$ for the tested and untested producers, respectively. At computer simu- lation, we set: $k_{tested} = 1$, $k_{untested} = 0.5$.

Then the j-th investor ranks all producers in accordance with the values A_{ij} and chooses the m most profitable producers with the large values A_{ij}. After this, the j-th investor forms the intention to distribute its total capital $K_{inv\ j}$ among the chosen producers proportionally to the values A_{ij}. Namely, the j-th investor intends to invest the capital C_{ij} into the i-th producer:

$$C_{ij} = K_{inv\ j}\frac{A_{ij}}{\sum_{i=1}^{M} A_{ij}}. \tag{7}$$

At the second iteration, each investor uses the intention agents to inform the selected producers about these values C_{ij}. Using this data, the producers evaluate their new expected capitals C'_{i0} in accordance with the expression (1).

Then the investors again send searching agents to all producers and estimate the new capitals of producers and the sums $\sum_{l=1}^{N} C_{il}$, taking into account the intentions of other investors. Profits of investors are evaluated by the expression (6), which already takes into account the intentions of all investors. Any investor ranks the producers and chooses the m most profitable producers again. After this, the investors estimate new planned values C_{ij} according to the expressions (6), (7). Once again, investors send intention agents to inform the producers about the planned capital investment values.

After a sufficiently large number of such iterations, the investors do the final decision about the investments for the next period T. Final capital investments are equal to the values C_{ij} obtained by the investors at the last iteration.

At the end of each period T, the capitals of producers are reduced to take into account the amortization processes: $K_{pro}(T + 1) = k_{amr} K_{pro}(T)$, where k_{amr} is the amortization factor ($0 < k_{amr} \leq 1$). The capitals of investors are reduced similarly (further, corresponding indicators are called inflation factors for convenience): $K_{inv}(T + 1) = k_{inf} K_{inv}(T)$, where k_{inf} is the inflation factor ($0 < k_{inf} \leq 1$).

3   Results of Computer Simulation

The described model was investigated by means of computer simulation. The simulation parameters were as follows:

- the total number of periods of considered processes: $N_T = 100$ or 500,
- the number of iterations in each period: $k_{iter} = 1,...,50$,
- the maximal thresholds of capitals of investors or producers (exceeding these thresholds leads to the reduplication of the investor or producer): $Th_{max\_inv} = 1$, $Th_{max\_pro} = 1$,
- the minimal thresholds of capitals of investors or producers (if the capital falls below these thresholds, then the corresponding investor or producer dies): $Th_{min\_inv} = 0.01$, $Th_{min\_pro} = 0.01$,
- the maximal number of producers and investors: $N_{pro\_max} = 100$, $N_{inv\_max} = 100$,
- the initial number of producers and investors: $N_{pro\_initial} = 2$ or 100, $N_{inv\_initial} = 50$ or 100,
- the maximal number of producers m, in which the investor can invest its capital, usually $m = 2$ or 100,
- the part of the profit that is transferred to investors: $k_{repay} = 0.6$,
- the characteristic variation of the coefficients k_i: $\Delta k = 0.01$,
- the parameters of function $F(x)$: $a = 0.1$, $Th = 100$.

The initial values k_i were uniformly distributed in the interval $[0,1]$.

The Specifics of the Iterative Process. In order to demonstrate the specifics of the iterative process clearly, we consider the results for the case of 2 producers and 50 investors. We assume that initial capitals of both producers are equal to 0.25 units. The production efficiencies k_i of the first and the second producers are equal to 0.5 and 0.9, respectively.

The first producer is tested ($k_{dist} = 1$), and the second producer is untested ($k_{dist} = 0.5$). Figure 1 presents the simulation results for the investor with the number one.

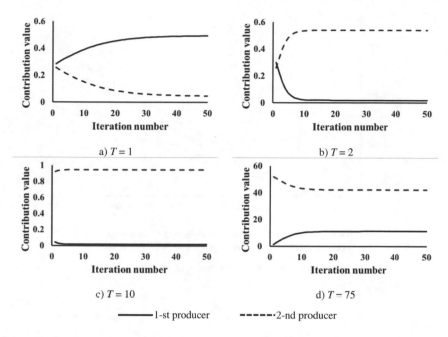

a) $T = 1$ b) $T = 2$

c) $T = 10$ d) $T = 75$

——— 1-st producer   - - - - - 2-nd producer

Fig. 1. The dependence of first investor contributions on the number of iteration at different periods T

The results characterize the following. At $T = 1$, when the more efficient second producer has not been tested, the investor from iteration to iteration increases the contribution to the first producer, despite its smaller efficiency (Fig. 1a). In the next period $T = 2$, the investor prefers the second more efficient producer (already tested), and the contribution to the first producer is gradually reduced (Fig. 1b). During the next periods, the investor contributes almost the entire capital into the second efficient producer (Fig. 1c). The investor makes such choice as long as the function $F(x)$ for the second producer does not reach the limit Th (see the expression (3)). After that, the investor begins to make a contribution to the first producer (Fig. 1d). Thus, it is beneficial to investors to make contributions into *perspective producers*, namely, into such producers, whose profits will grow with increase of their capital. The iterations play the important role in these processes of adjustment of contributions.

The Effectiveness of Iterative Evaluations for the Case N = M = 100. In order to show that investors are more successful, if they take into account the intentions of other investors, we simulate the processes without the iterative estimates ($k_{iter} = 1$) and with iterations ($k_{iter} = 50$). We consider two cases: (1) without amortization and inflation and (2) with amortization and inflation. Figure 2 demonstrates that the iterations increase the capital of both investors and producers. Without amortization and inflation, the iterations increase the capital of the community by 10% (Fig. 2a). In the case of amortization and inflation, the effect is more significant, the iterations increase the capital of producers and investors by 41–43% (Fig. 2b).

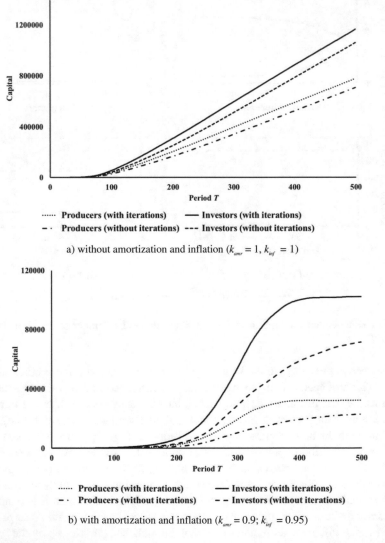

a) without amortization and inflation ($k_{amr} = 1$, $k_{inf} = 1$)

b) with amortization and inflation ($k_{amr} = 0.9$; $k_{inf} = 0.95$)

Fig. 2. Influence of iterative evaluations. The dependence of the total capital of the producers and investors on period T

4   Conclusion

Thus, the processes of self-organization in the community of producers and investors have been analyzed. Original features of the current model are the following: (1) the cooperation between investors and producers, (2) the openness of information about the current capitals and effectiveness of the producers and about the intentions of investors to invest capitals into different producers, (3) the iterative process of the formation of capital investments. The most important result of the model is the development of the new method for profitable capital investments. It is beneficial to investors to make contributions into *perspective producers*, namely, into such producers, whose profits will grow with increase of their capital.

Acknowledgments. This work was partially supported by the Russian Foundation for Basic Research, Grant No 16-01-00223.

References

1. Axelrod, R.: The Evolution of Cooperation. Basic Books, New York (1984)
2. Burtsev, M., Turchin, P.: Evolution of cooperative strategies from first principles. Nature **440**(7087), 1041–1044 (2006)
3. Claes, R., Holvoet, T., Weyns, D.: A decentralized approach for anticipatory vehicle routing using delegate multiagent systems. IEEE Trans. Intell. Transp. Syst. **12**(2), 364–373 (2001)
4. Holvoet, T., Valckenaers, P.: Exploiting the environment for coordinating agent intentions. In: Environments for Multi-Agent Systems. Lecture Notes in Artificial Intelligence, Part III, vol. 4389, pp. 51–66 (2007)

Neurobiology

Complexity of Heart Rate During More and Less Differentiated Behaviors

Anastasiia Bakhchina[✉]

Institute of Psychology, Russian Academy of Sciences, Moscow, Russia
nastya18-90@mail.ru

Abstract. Autonomic nervous system is the main way for the brain–body coordination, of which mode can be evaluated by dynamics of heart rate variability (HRV). HRV analysis is used for evaluation of different psychological states (stress, arousal, cognitive control etc.), which can be considered as characteristics of behavior that formed at different stages of ontogeny. We investigated whether HRV differs between the early-formed (less differentiated) behavior and the later-formed (more differentiated) behavior. Heart rate was recorded in 33 healthy subjects (mathematical specialists). Participants performed two tests which included sentences with mathematical terms and sentences with common current used words. They had to add one missing word in each sentence. Sample entropy as a measure quantifying the complexity of time series was used to analyze HRV. Complexity of heart rate was significantly higher in the mathematical test performance when participants actualized the later-formed behavior.

Keywords: Autonomic nervous system · Complexity of heart rate · Different stages of ontogeny · Behavioral complexity · Sample entropy

1 Introduction

Physiological supporting of behavior includes activation of different linked neurons groups and optimization of physiological processes. As a rule traditionally in the conserved approach internal bodily states are ignored in the searching of the neural basis of behavior. Such mental functions as perceptions, thoughts, feelings etc. are for the most part considered in isolation from the physiological state of the body. A mechanistic understanding of distinct interoceptive pathways, which can influence brain functions, leads to the impossibility of considering the behavior at the whole organism level [1]. Therefore the current main task is forming of the system approach for describing of behavior from psychophysiology perspective.

Autonomic nervous system takes part in subserving of behavior. Studies of autonomic psychophysiology are beginning to have a big part in the current field of neuroscience. As an example, the fundamental association between bodily changes and emotions was founded by James and Lange at 19 century. Since that time a lot off studies with electrical stimulation in animals have demonstrated the coupling of visceral responses to cortical regions, which include cingulate, insular [2], visual [3] and

B. Kryzhanovsky et al. (eds.), *Advances in Neural Computation, Machine Learning, and Cognitive Research*, Studies in Computational Intelligence 736,
DOI 10.1007/978-3-319-66604-4_25

somatosensory [4] regions. It means that not only the nucleus of the solitary tract, ventrolateral medulla, parabrachial nucleus and hypothalamus but also many cortex regions take part in the brain–body cooperation (processing of visceral information). Experimental researches into the mechanisms through which visceral afferent information is represented within the brain haven't shown clearly how visceral signals shape human behavior yet. The majority of visceral signals that shape behavior are unnoticed despite there is anatomical and experimental information about the representation and influence of the visceral state in brain processes.

In this way describing principles through which human behavior and experience is coloured by internal bodily signals Benarroch [5] showed the central autonomic network (CAN), which included different structures of the central and the autonomic nervous system. The main statement of the theory is the CAN is an integral component of an internal regulation system through which the brain controls visceromotor, neuroendocrine, pain, and behavioral responses essential for survival (for goal-directed behavior supporting).

Heart rate variability can be considered as a tool for measurement of autonomic nervous system activity. Heart rate variability (HRV) is the variation over time of heart beat intervals (the periods between consecutive heartbeats), which depends on such physiological processes as autonomic neural regulation, thermoregulation, breathing etc. [6]. HRV is thought to reflect the heart's ability to adapt to changing behaviour and can be considered as an indicator of central-peripheral neural feedback and central nervous system – autonomic nervous system integration. Therefore HRV was used in the current study as a noninvasive tool for assessing the activities of the autonomic nervous system.

It is shown that HRV associated with a diverse range of processes, including affective and attention regulation, cognitive functions (such as working memory, sustained attention, behavioral inhibition, general mental flexibility) [7]. These processes can be considered as characteristics of behavior formed at different stages of ontogeny.

From the system-evolutionary theory [8] perspective, a new behavior is sub-served by co-activation of specialized neurons groups that had emerged in learning. The result of learning is a functional system that is a set of brain and body elements activity for providing efficient interaction with the environment [9]. Formation of new systems during development results in growing complexity and differentiation of organism–environment relations. They are becoming more detailed and specific. Consequently, the structure of behavior becomes more complex and differentiated during development [10]. In this way ontogenetic development can be considered as the process of increasing differentiation along with the number of learnt behavioral functional systems [11].

Therefore we investigated whether HRV in the early-formed behavior ("old") differs from HRV in the later-formed ("new") behavior. Basing on the fact that usually "old" behavior is less complicated than "new" behavior [12] we hypothesized that heart rate complexity would be higher at "new" behavior performing.

2   Materials and Methods

The experiment was approved by the Ethics Committee of the Institute of Psychology of Russian Academy of Science. Prior to the experiment, all subjects signed an informed consent form stating that participation was voluntary and that they could withdraw from the study at any moment.

Thirty-three healthy subjects (28 men, median = 27.78 years, range 23–37 years) participated.

Participants did not suffer from any self-reported respiratory, cardiac diseases, epilepsy, psychiatric disorder, or any minor or major illness. All participants were professional mathematicians and had work experience (Median = 4.84 years).

The linguistic task was used for modeling behaviors formed at different stages of ontogeny because language acquisition, as part of individual development, can be considered as learning to achieve adaptive results [13, 14].

The experimental task was to add one missing word in the sentence. The sentences (N = 64) were divided into 2 groups by age of acquisition of words. The first group of sentences (N = 32) included sentences with mathematical terms. Subjects had known these terms from University studying (from the age of 18–19 years). The example is "A normal is a vector that is perpendicular to a given object". These sentences included later-acquired words and made actual a "new" behavior. The second group of sentences (N = 32) included sentences with commonly used words. Subjects had known these words from childhood (from the age of 5–6 years). The example is "Plasticine is a material for modeling figures". These sentences included early-acquired words and made actual an "old" behavior. The sentences in both groups were equal in the linguistic estimations, such as the quantity of words, syllables, letters and Fog's index.

The sentences were performed individually on a standard computer. The order of sentences was random without repetition in each group.

The ECG was obtained using the wireless device HxM BT by Zephyr Technology and the developed software complex. The plastic electrodes were filled with electrolyte and placed on the thorax across the heart, they were located in I and II chest leads. Batch data transmission from the sensor to the mobile device was done through the wireless protocol Bluetooth. Realization of communication, data transmission and storage was performed in the mobile device by the original software "HR-Reader" [15]. "HR-Reader" program medium provided on-line visualization of the registered signal for the record control. The signal was sampled at 400 Hz. The inter-beat intervals (IBI) were extracted from ECG through the threshold algorithm.

The time domain indexes of HRV used in the analysis were the mean (HR, ms) and standard deviation (SDNN, ms) of IBI. These indexes closely reflect all nervous regulatory inputs to the heart.

For estimation of heart rate complexity we used the sample entropy (SampEn) as a set of measures of system complexity reporting on similarity in time series. SampEn was chosen because it is successfully applied to relatively short and noisy data and it is largely independent of record length and displays relative consistency under circumstances. SampEn (m, r, N) is precisely the negative natural logarithm of the conditional probability that two sequences similar for m points remain similar at the next point, where self-matches are not included in calculating the probability [16]. The parameter N is the length of the time series, m is the length of sequences to be compared; r is the tolerance for accepting matches. Thus a low value of SampEn reflects a high degree of regularity. SampEn is independent on the record length and displays relative consistency under circumstances. The parameters m and r were fixed: m = 2, r = 0.5*SDNN.

We calculated HR, SDNN and SampEn for series of IBI individually during performance of the task with mathematical words and the task with commonly used words.

Normality of variables was tested in Shapiro–Wilk's test (W-test). HRV data of two conditions (performing tasks with mathematical or commonly used words) were tested in Wilcoxon signed-rank test. We used non-parametric test because the majority of variables didn't have the normal distribution. Statistical analyses for all measures were accomplished with Statistica 10.

3   Results

We compared to time domain indexes (HR and SDNN) and non-linear index (SampEn) of HRV between two periods: performing the task with mathematical words and performing the task with commonly used words, using non-parametric Wilcoxon signed rank test.

Values of heart rate (HR) and of the standard deviation of heart rate (SDNN) did not significantly differ between two conditions (Tables 1 and 2).

Table 1. Description statistics (median, lower and upper quartiles) and the results of Shapiro–Wilk's test of HRV parameters in task performing with sentences with mathematical words (MW) and sentences with commonly used words (CW).

Statistics	SamEn CW	SampEn MW	SDNN CW	SDNN MW	HR CW	HR MW
Median	0.65	0.72	56.72	58.92	789.64	781.35
Lower quartile	0.51	0.62	44.67	45.48	734.11	708.86
Upper quartile	0.77	0.79	76.48	76.88	922.31	907.13
Median	0.65	0.72	56.72	58.92	789.64	781.35

Table 2. The distributions of the medians of HRV parameters in two types tasks performing were compared using Wilcoxon signed-rank test.

Variables	T	Z	p
SampEn	82.00	2.37	0.01*
SDNN	128.00	1.21	0.22
HR	132.00	0.65	0.49

*Significant level.

Heart rate complexity (SampEn) was significantly higher in the performing the task with mathematical terms than performing the task with commonly used words (Tables 1 and 2).

4   Conclusion

The aim of the current study was to examine the relationship between the system supporting of behavior and complexity of heart rate. It was shown that the early-formed behavior, which realized less differentiated organism–environment relationships, was corresponded with less complexity of heart rate than the latter-formed behavior, which realized more differentiated organism–environment relations.

It was shown that neuronal subserving of the latter-formed behavior includes more neuronal systems [17]. As an example, the acute effect of alcohol on the ERPs related to the use of knowledge and experiences acquired at the early stages of individual development was less than at the later stages [18]. Therefore we can suppose that during later formed behavior the central-autonomic network has to realize more nonstationary activity which demands many different changes in the activity of the heart and other parts of the organism. It leads to lack of regularity of heart rate and to the increase of complexity (Fig. 1).

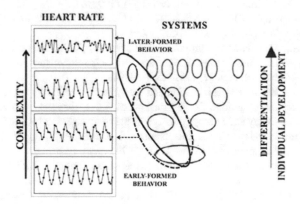

Fig. 1. The correlation between complexity of heart rate and degree of differentiation of behaviors formed at different stages of individual development. The small ovals depict functional systems formed at different stages of individual development. Groups of ovals connected by the line illustrate the combination of functional systems that provide realization of behavior: full line - later-formed behavior, dashed line - early-formed behavior.

It is important that HR was the same in both conditions. It means that these different modes of heart activity, which were seen in the results, cannot be explained through the different intensity of cognitive load, which demands different quantity of internal recourses during early- and latter-formed behaviors.

The main output of the study is that the system subserving of behavior is reflected not only in the brain activity but also in the body activity. Functional systems, which subserve behavior, are not only neuron systems. They also include different parts of the body, which change their activity in cooperation with the brain for an optimal achievement of results.

Acknowledgments. The search is supported by grant RFBR N16-36-60044 mol_a_dk, within the research programme of a Leading Scientific School of Russian Federation: "System Psychophysiology" (NSh-9808.2016.6).

References

1. Critchley, H.D., Harrison, N.A.: Visceral influences on brain and behavior. Neuron **77**, 624–638 (2013)
2. Bagaev, V., Aleksandrov, V.: Visceral-related area in the rat insular cortex. Auton. Neurosci. Basic Clin. **125**, 16–21 (2006)
3. Pigarev, I.N., Bagaev, V.A., Levichkina, E.V., Fedorov, G.O., Busigina, I.I.: Cortical visual areas process intestinal information during slow-wave sleep. Neurogastroenterol. Motil. **25**, e268–e169 (2013)
4. Brüggemann, J., Shi, T., Apkarian, A.V.: Viscero-somatic neurons in the primary somatosensory cortex (SI) of the squirrel monkey. Brain Res. **756**, 297–300 (1997)
5. Benarroch, E.E.: The central autonomic network—functional organization, dysfunction, and perspective. Mayo Clinic Proc. **68**(10), 988–1001 (1993)
6. Acharya, U.R., Joseph, K.P., Choo, N.K., Jasjit, M.L., Suri, S.: Heart rate variability: a review. Med. Biol. Eng. Comput. **44**, 1031–1051 (2006)
7. Matthews, S.C., Paulus, M.P., Simmons, A.N., Nelesen, R.A., Dimsdale, J.E.: Functional subdivisions within anterior cingulate cortex and their relationship to autonomic nervous system function. NeuroImage **22**, 1151–1156 (2004)
8. Shvyrkov, V.B.: Behavior specialization and the system-selection hypothesis of learning. In: Human Memory and Cognitive Capabilities, pp. 599–611. Elsevier, Amsterdam (1986)
9. Anokhin, P.K.: Biology and neurophysiology of the conditioned reflex and its role in adaptive behavior: scientific and translation. In: Corson, S.A., Dartau, P.-R., Epp, J., Kirilcuk, V. (eds.) (Transl.), 98 pp. Pergamon Press, New York (1974)
10. Edelman, G.M.: Naturalizing consciousness: a theoretical framework. PNAS **100**, 5520–5524 (2003)
11. Alexandrov, Y.I.: Physiological Significance of the Activity of Central and Peripheral Neurons in Behavior, 426 pp. Science, Moscow (1989)
12. Lewin, K.: Action research and minority problems. In: Lewin, G.W. (ed.) Resolving Social Conflicts, 412 pp. Harper & Row, New York (1946)
13. Maturana, H.R.: Biology of cognition. BCL Report No. 90, pp. 5–58. University of Illinois, Urbana (1970)
14. Kolbeneva, M.G., Alexandrov, Y.I.: Mental reactivation and pleasantness judgment of experience related to vision, hearing, skin sensations, taste and olfaction. PLoS ONE **11**(7), e0159036 (2016). doi:10.1371/journal.pone.0159036
15. Runova, E.V., Grigoreva, V.N., Bakhchina, A.V., Parin, S.B., Shishalov, I.S., Kozhevnikov, V.V., Necrasova, M.M., Karatushina, D.I., Grigorieva, K.A., Polevaya, S.A.: Vegetative correlates of conscious representation of emotional stress. CTM **4**, 69–77 (2013)
16. Richman, J.S., Moorman, J.R.: Physiological time-series analysis using approximate entropy and sample entropy. Am. J. Physiol. Heart Circ. Physiol. **278**, H2039–H2049 (2000)

17. Yu, I.A., Grinchenko, V.G., Shevchenko, D.G., Averkin, R.G., Matz, V.N., Laukka, S., Korpusova, A.V.: A subset of cingulate cortical neurons is specifically activated during alcohol-acquisition behavior. Acta Physiol. Scand. **171**, 87–97 (2001)
18. Yu, I.A., Sams, M., Lavikainen, J., Reinikainen, K., Naatanen, R.: Differential effects of alcohol on the cortical processing of foreign and native language. Int. J. Psychophysiol. **28**, 1–10 (1998)

Comparison of Some Fractal Analysis Methods for Studying the Spontaneous Activity in Medullar Auditory Units

N.G. Bibikov[(✉)] and I.V. Makushevich

JSC NN Andreev Acoustical Institute, Moscow, Russia
nbibikov1@yandex.ru, iliamail@rambler.ru

Abstract. We recorded the spontaneous background impulse activity in the medullar and midbrain single auditory neurons of the paralyzed grass frog. This activity was considered by us as a chaotic point process. For the analyses of temporal changes of this process we used the approach based upon the recording of the Hurst index. This approach was juxtaposed with the methods based on the study of the dependence of Fano and Allan factors on the duration of the analyzed interval. A comparative analysis of the Fano, Allan and Hurst indices by Kendall's rank correlation method have been made. We observed a close correlation of the values of the Fano and Allan indices for the spontaneous activity of the same neuron. Correlation of the Fano and Hurst indices was not so pronounced and did not quite correspond to the properties of typical fractal point processes. It is possible to formulate assumptions about the possibility and efficiency of using Hurst index to analyze the sequence of pulsed discharges of a neuron. In most cells, chaotic changes in impulse density were observed, which is indicative of the trend behavior of neuron's firing. Anti-trend behavior was not observed.

Keywords: Amphibians · Spontaneous activity · Fractal · Allan factor · Fano factor · Hurst index

1 Introduction

The spontaneous (background) activity of auditory neurons generated in the absence of controlled external sounds reveals many important properties of test cells, including both dynamics of the recovery of excitability after generation of spike discharge, and relatively slow changes of the firing density, that reflect important processes occurring in brain neural networks. In the Russian Acoustics Institute the study of background activity of frog auditory units had started a long time ago [1], and continues up to now.

During this period we have used several methods for analysis of the point process of background neuronal firing, including fractal approach [2–5]. The results of the analysis of spontaneous activity of the auditory neurons of the medulla oblongata of the frog using the Fano and Allan factors were presented in 2009 [6, 7]. The use of these techniques has allowed us to arrive to a number of conclusions. For time intervals of less than one second, the background activity of the majority of units could be adequately

© Springer International Publishing AG 2018
B. Kryzhanovsky et al. (eds.), *Advances in Neural Computation, Machine Learning, and Cognitive Research*, Studies in Computational Intelligence 736,
DOI 10.1007/978-3-319-66604-4_26

described as random process, although some neurons reveal a marked drop in the values of the Fano and Allan factors. This effect is likely determined by the presence of the accumulating refractory. However, for the duration of the periods longer than one second, most neurons demonstrated the power function growth of the Fano factor values, indicating chaotic and fractal properties of the investigated process. The growth of Allan factor values usually occur only at longer duration of the analyzed periods.

This report presents the results of processing of the spontaneous firing in medullar neurons of the frog using Hurst index, which was initially introduced for the analysis of changes of the Nile's water level [4]. Later it was used for the analysis of a variety of processes, both natural and social. We compare results obtained by Hurst index with the results of analysis of neuronal activity using the Fano and Allan factors [6, 7].

2   Methods

2.1   Electrophysiological Recording of the Background Firing

The raw data on the impulse activity of single neurons in the medulla oblongata were obtained in electrophysiological experiments on amphibians (immobilized frogs - Rana t temporaria). Extracellular recordings were carried out in the dorsal medulla nucleus using glass electrodes filled by NaCl with resistance of 2–10 mgOm and an amplifier with high input impedance. During the operation complied with the requirements of humanitarian treatment of animals we have used MS222 or cold anesthesia. During recordings the frog was placed in a soundproofed booth with minimum ambient noise. Registration was carried out in the dorsal medullar nucleus whose neurons receive inputs from auditory nerve fibers. Identified spike pulses were converted via Schmitt trigger into standard electric pulses stored in the memory of a personal computer. Time of occurrence of each spike was recorded with a precision of 0.5 ms. Experimental methods were described in detail previously [6, 8].

2.2   Data Processing

We used the standard formula for the calculations of the functions characterizing the dependence of Fano and Allan factors on the interval duration.

Fano factor:

$$F(T) = \frac{D[N_i(T)]}{E[N_i(T)]} = \frac{\sigma^2(T)}{\mu(T)} \tag{1}$$

Allan factor:

$$A(T) = \frac{\left\langle N_{i+1}(T) - N_i(T) \right\rangle^2}{2\langle N_i(T) \rangle} \tag{2}$$

where T is duration of the time interval, N is the number of spikes in analyzed interval.

The values of exponents of these functions in the intervals where they were close to a straight line in the double logarithmic coordinates are called here Fano (or Allan) indexes. The intervals were considered suitable for indexes calculation, if they included at least five points and the correlation coefficient of the dependence was greater than 0.9. Since such intervals were absent in some units, these factors was calculated not for all investigated neurons.

Figure 1 shows examples of the dependences of the Fano factor values upon the length of the analyzed intervals. The first of the illustrated neurons exhibits typical behavior (Fig. 1a). However, in some cases, this dependence is almost absent for all analyzed intervals (Fig. 1b).

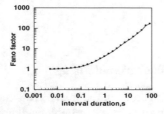 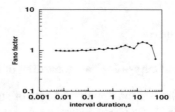

Fig. 1. Typical plots of the "power" (a) and "non-power" (b) behavior of the Fano factor in double logarithmic coordinates.

The method of the Hurst index calculation, which we used in our study, was taken from the paper [9].

The formula for computing the Hurst index:

$$H = \frac{\log\left(\dfrac{R}{S}\right)}{\log\left(\dfrac{1}{2}n\right)} \tag{3}$$

where: n is the number of measurements (members of the time series), S is the standard deviation of a series of observations, and R is the difference between maximal and minimal of the time series intervals.

In some papers, for the analysis of point temporal sequence several methods are applied simultaneously to the analysis of fractal processes. For example, in the paper [10] for a specific object - the sensor network router - the DFA method, correlation analysis, Fano factor and Hurst index were applied.

In our study, we carried out a comparative analysis of the Hurst index with Fano and Allan indexes for neurons, which gave information on all these factors. For the "case table" thus formed, the Kendall rank correlation coefficient was calculated [11].

The values of the Kendall coefficient were calculated for a "complete" base of neurons and for groups of neurons with a successively decreasing number of spikes. Our purpose was to check the interrelations of the Hurst index with indexes of Fano and Allan.

Table 1 shows the results of the Hurst index calculation from the raw data. In the majority of cases, the Hurst index significantly exceeds the value of 0.5, which is critical for random behavior. There are several neurons for which the Hurst index is close to 0.5. We did not observed a single neuron with the Hurst index less than 0.5.

Table 1. Hurst index values

№ spikes	N cells	Mean	Median	Mode
All cases	60	0.7362	0.72	0.70
N ≥ 5000	29	0.7524	0.75	0.79 and 0.83
N ≤ 5000	31	0.7210	0.70	0.70
N ≤ 4000	24	0.7138	0.70	0.70
N ≤ 3000	18	0.7272	0.71	0.70 and 0.71
N ≤ 2000	7	0.7257	0.67	0.62

We analyzed the relationship between the Hurst, Allan and Fano indexes using Kandall's rank correlation coefficients (Tables 2 and 3). From the data in the Table 2, we can conclude that there is a statistically significant correlation between Fano and Allan indexes for all values of N (the number of spikes), since the value of significance level (p) has always been less than 0.05.

Table 2. Kendall's rank correlation coefficients between Fano and Allan indexes

N spikes	N cells	Kendall coefficient	Significance
All cases	60	0.490	0.000
N ≥ 5000	29	0.575	0.000
N ≤ 5000	31	0.446	0.000
N ≤ 4000	24	0.352	0.002
N ≤ 3000	18	0.311	0.042
N ≤ 2000	7	0.524	0.004

Table 3. Kendall's rank correlation (Krc) between the Hurst, Fano and Allan indexes

Number of spikes	Number of cells	Hurst index vs Fano factor		Hurst index vs Allan factor	
		Krc	p	Krc	p
All cases	60	0.408	0.000	0.344	0.000
N ≥ 5000	29	0.498	0.000	0.451	0.000
N ≤ 5000	31	0.346	0.010	0.287	0.027
N ≤ 4000	24	0.156	0.368	0.117	0.505
N ≤ 3000	18	0.265	0.134	0.114	0.564
N ≤ 2000	7	0.238	0.571	0.143	0.722

Table 3 shows the coefficients of the relationship between the Hurst index and the Fano and Allan indexes. For the entire experimental base of 60 neurons, a statistically significant correlation was recorded. However, this dependence is fairly clear only when the sample was very representative.

For N < 4000 values, we could not find statistically significant association. A graphical comparison of the Fano the Hurst indexes is given in Fig. 2. The linear approximation was described by the formula $H = 0.192 * F + 0.621$ (dotted line). The theoretical line $H = 0.5 * F + 0.5$ (solid line), which assumes a purely fractal nature of the processes is also shown. Although the experimental data are in qualitative agreement with the theoretical dependence, the relationship of these parameters is only qualitative.

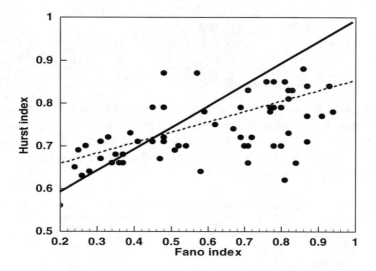

Fig. 2. The relationship between the Fano and the Hurst indexes.

In the paper [9] a modified Hurst index was introduced and a technique for classifying time series as random or trend/anti-trend for time series with $N \leq 5000$ was developed (based on the author's calculations using pseudorandom data). This approach, in principle, allows us to follow the dynamic behavior of the Hurst index on individual successive segments of the process. Preliminary results suggest that the behavior of the process describing the spontaneous activity of the auditory neurons of the frog's medulla can spontaneously vary from trend to random and backward.

3   Conclusion

A comparative analysis of the relationships between Fano, Allan and Hurst indexes using the Kendall correlation coefficient demonstrates that all these parameters can be used for statistical description of spontaneous neuronal activity. However, a statistically significant relationship between the Hurst index, on the one hand, and the of Fano and Allan indexes, on the other, was found only for representative samples (with a number of spikes $N > 4000$).

The work was supported by the RFBR grant 16-04-01066. The authors are deeply grateful to S.V. Nizamov and A.A. Pyrogova for help with software.

References

1. Bibikov, N.G.: Investigation of spontaneous activity of neurons in the acoustic center of the frog's midbrain. Statistical Electrophysiology. Vilnius: Science, pp. 70–80 (1968). (in Russian)
2. Mandelbrot, B.: Fractal Geometry of Nature/Translation from English by A.R. Logunov, 656 pp. Institute for Computer Research, Moscow (2002). (in Russian)
3. Fano, U.: Ionization yield of radiations. II. Fluctuations Number Ions. Phys. Rev. **72**(1), 26–29 (1947)
4. Hurst, H.E.: Long term storage capacity of reservoirs. Trans. Am. Soc. Civ. Eng. **116**(1), 770–799 (1951)
5. Barnes, J.A., Allan, D.W.: A statistical model of Flicker noise. Proc. IEEE **54**(2), 176–178 (1966)
6. Bibikov, N.G., Dymov, A.B.: Fano and Allan factors of the process of spontaneous impulse activity of acoustic neurons of the medulla oblongata. Sensor. Syst. **23**(3), 246–259 (2009). (in Russian)
7. Dymov, A.B.: Use of Fano and Allan factors to analyze the properties of the spike sequence of neurons in the auditory system. In: Proceedings of the XI All-Russian Scientific and Technical Conference "Neuroinformatics-2009", pp. 257–263. MIFI Press, Moscow (2009) (in Russian)
8. Bibikov, N.G.: Correlation of the responses of the neurons in the cochlear nucleus of the frog with low-frequency noise amplitude modulation of the tone signal. Acoust. Phys. **60**(5), 597–607 (2014)
9. Naiman, E.L.: Calculation of the Hurst index for the purpose of revealing the trend (persistence) of financial markets and macroeconomic indicators. http://wealth-lab.net/Data/Sites/1/SharedFiles/doc/forindicators/articles/04_erik_naiman_herst.pdf
10. Aksyonov, V.Y., Dmitriev, V.N.: Algorithms of fractal analysis of time series in monitoring systems of sensor networks. Bull. Astrakhan State Tech. Univ. Ser. Manag. Comput. Sci. Inform. **1**, 91–96 (2012). (in Russian)
11. Kendall, M., Stewart, A.: Statistical Conclusions and Connections, pp. 687–718. Nauka, Moscow (1973)

Synapse as a Multi-component and Multi-level Information System

A.L. Proskura, A.S. Ratushnyak, S.O. Vechkapova,
and T.A. Zapara$^{(\boxtimes)}$

Institute of Computational Technologies,
Siberian Branch of the Russian Academy of Sciences, Novosibirsk, Russia
Zapara_t@mail.ru

Abstract. Synapse, as it is known, consists of presynaptic and postsynaptic parts. In many brain cells, particularly in the cortex and the hippocampus, a postsynaptic part presents a small membrane protrusion on the surface of the dendrite - a dendritic spine. The dendritic spine is partly an isolated structure, but it is associated functionally with other spines of dendritic shaft as well as through the vesicular system with the soma of neuron. The main task of the spine interactome is not only to receive a signal from a presynaptic cell, but also to react to it by opening/closing the ion channels, thus ensuring its transmission to the axon. The interactome of spine is primarily a detector of environmental signals and through the remodeling of the system of its macro-complexes it recognizes and remembers the pattern of the signal.

Keywords: Pyramidal neuron · Synapse · Interactome · Modulation · AMPA – and NMDA-type receptors

1 Introduction

A synapse is the place of the contact between neurons or between a neuron and a signal-receiving effectors cell. It serves to transmit a nerve impulse between two cells, and during the synaptic transmission, the amplitude and frequency of the signal can be regulated. The interactome of dendritic spine (a set of all protein interactions with each other) represents a dynamic multi-level and multi-component system.

The evolution of the brain correlates with complication of a molecular organization of nerve cell [1–3]. The complexity of neuron organization allows considering it as the organism in the organism. Sherrington [4] proposed the concept of the cell as an organism in the organism. The nerve cell, like any living cell, implements a genetic program providing one's "demand/motivation" owing of metabolism. The neurons cannot procure themselves metabolites from environment using own mobility as free organisms. The neurons receive metabolites from the other cells and must integrate with the cells of the organism in functional systems supporting their metabolism.

The perception–action cycle is a basic biological principle that governs the functional relationships of the organism with its environment and guides the organism to its goals [5, 6].

B. Kryzhanovsky et al. (eds.), *Advances in Neural Computation, Machine Learning, and Cognitive Research*, Studies in Computational Intelligence 736,
DOI 10.1007/978-3-319-66604-4_27

In recent years, the paradigm of a distributed network of cortical memory has emerged, and the concept of a memory network [7] which are the center of new paradigms of the memory. A memory or an item of knowledge is defined by a pattern of connections between neuron populations associated by experience and a neuron can be part of many memory networks and, thus, element of knowledge [7]. The neuron is also interacting with other neurons in the systems to provide the specific functions of the brain. It is obvious that the activity of the neuron, as element of memory network, as the systems of provision by metabolites is the actions that lead to the achievement of the aim by function systems of cell or brain networks. The functional systems within a neuron are the molecular networks which are compartments performing the specific cellular function. The activity of the neuron entails changes in their connections with other neurons as well in their molecular network.

Learning and memory are mediated by changes in the excitability of the cell and modulations of synaptic transmission, usually it take place in protein system of the postsynaptic (dendritic) part of a synapse. The hippocampus is a brain structure involved in short-term memory and the phenomenon of long-term potentiation (LTP) (increasing the efficiency of synaptic transmission after an intense and short-term release of neurotransmitter) is the dominant cellular model of learning and memory [7].

The implementation of a neuron function depends on the actions of its numerous molecular-effector systems. The generalization of the complex processes of emergence synaptic memory that occur even in separate neuronal compartments without special tools is a difficult, if at all possible, task. The technology which combines the creation of a database with presentation them in the form of networks facilitates this process. The protein-protein interaction network in dendritic spines of the hippocampal pyramidal neurons facilitates the synthesis of numerous experimental data in conceptual knowledge about the principles and molecular mechanisms of operation the neurons.

The main purpose of this review is to reconstruct the action sequence of functional (executive) systems of the neuron involved in synaptic modulations in the phases of long-term potentiation.

2 The Own Goals of the Individual Neuron

The individual neuron of multi-cellular biological systems has cognitive functions. Neurons isolated in vitro display the cognitive-like "behavior" - remembering and recognizing signals, predicting the possible changes in the external environment, and choosing the appropriate reaction to prevent their consequences [8]. For the neuron, brain is its environment. The survival of the individual neuron is based on genetic programs adapted to the conditions of environment by learning in the process of life. Homeostasis is a major genetically predetermined program and it is maintained by motivational subprograms. Evaluation and maintenance of energy resources (food motivation), avoiding the adverse conditions (defensive motivation); the reparation of damages, generation of new processes (information input) and algorithms of reactions on the basis of association of signals, and prediction of future signals are the major neuron subprograms [9–11]. Comparing the signals generated by an intracellular molecular information system with data from memory (multi-level, distributed in each of these systems and in the genome) with the signals coming from the outside medium (for the cell) leads to the classification of a certain degree of novelty of these signals. If a signal is recognized, then the system predicting the consequences acts on the basis of algorithms that genetically determined and gained during learning.

The molecular basis of this multilevel functional system is likely the intracellular molecular assemblies, with the cytoskeleton acting as the substrate for associative memory. The dipole system of the tubulin cytoskeleton and clusters of other polarized proteins performs the pattern recognition [12–14], thereby controlling the effectors. The activation of cellular effectors leads to the formation of a response to avoid or minimize external influences by changes in both the intracellular system and the environment. The processes causing the formation of intercellular connections and synaptic modulations for a particular neuron are changes in the external environment.

The functional systems within a neuron are the molecular networks of compartments performing specific cellular functions. The interaction of these systems leads to the creation of new integrative properties (the so-called emergent-systemic properties and functions), that are not the part of the individual components. These newly emerged properties allow the nerve cells to function as the complex of molecular information systems that underlie cognitive functions.

Nowadays the development of a network of executive systems that guides the neuron to its numerous parallel goals is an urgent task. The modern level of conceptual knowledge and considerable amount of experimental data does not allow creating the complete network. The functional systems [7], the autonomous adaptive agents [15] are circuits that the most comply with the experimental data. We developed a structure-functional schema of a neuron as a system with elements of cognition functions.

3   The Functional Systems of the Neuron Involved in Synaptic Modulations in the Early Phase of LTP

Dendritic spines are the postsynaptic component of a synapse. These protrusions (less than 1 μm^3) originate from an axial dendrite. Networks of the dendritic spine proteins of pyramidal neurons of the CA1 hippocampal region of rodent in the early phase of LTP have been reconstructed using GeneNet computer system [16]. The protein-protein interaction (PI1) network of dendritic spines is presented on the site (http://wwwmgs. bionet.nsc.ru/mgs/gnw/genenet/viewer/Earlylong-termpotentiation.html).

Network components and their relationships were developed by referring to the published articles (PubMed) and databases (Swiss-Prot, EMBL, MGI, GeneCard, and TRRD). In addition to the qualitative characteristics, such as presence or absence of links between proteins, the system takes into account the nature of the regulatory interactions between proteins, i.e., activation, inactivation, and enhancement or suppression of the molecular interactions. PI1 network reflects the organization of the system in one of the possible dendritic spine states. In constructing PI1 network, we considered that a spine is a highly ordered structure with a specific horizontal and vertical organization at all levels: the membrane (synaptic, perisinaptic extrasinaptic zone); submembrane (cytoplasmic tails of receptors and submembrane proteins); and the cytoplasm (the nanodomains of proteins, the network of actin filaments, intracellular stores of receptors, and other proteins).

Proteins that anchor the receptors and bind them to the cytoskeleton (scaffold proteins), protein kinases, proteases, and GTPases form the postsynaptic density (PSD) located under the synaptic membrane [17].

Glutamate receptors are receptor-channel complexes. The classification of these receptors is based on their sensitivity to N-methyl-D-aspartate and (NMDA) and α-amino-3-hydroxyl-5-methyl-4-isoxazole-propionate (AMPA).

MDARs are directly anchored at the center of the PSD and AMPAR are anchored through a family of stargazing proteins at the periphery [18, 19]. In PSDs, there are the physical interactions between molecules and positioning the partner molecules in complexes (macro-complex). AMPARs form a connection with 9 proteins in PSD, and NMDARs form a connection with more than 450 proteins in PSD [20, 21].

The dendritic spine cytoskeleton comprises a polymeric (F-actin) and monomeric (G-actin) actin, and F-actin is dominant.

Transduction of signals from receptors to the actin cytoskeleton is mediated by the small GTPase family, protein regulators of small GTPases, kinases, phosphatases, and numerous regulatory proteins that directly interact with G-actin and F-actin.

Small GTPases circulate between two their states: active - GTP-bound (guanine triphosphate) and inactive - GDP-bound (guanine diphosphate). For example, our network shows that the small GTPase Rac1 is activated by the GEF Tiam1Rac1 and inactivated by the GAP MEGAP Rac1, and is indirectly regulated through SynGAPRas. Thus, PSD scaffold proteins are the original sites for the assembly of functional nanodomains, providing a point of physical interaction between effectors, acceptors, and their activators and inhibitors [22].

PI1 network of dendritic spines reflects the molecular executive systems, which provides increasing the excitability of the cell after an intense and short-term release of neurotransmitter. This pattern of glutamate receptor activation leads to the transition of spine to the level, corresponding to more efficient synaptic transmission.

PI1 network of dendritic spines and the description of the processes that are initiated by activation of glutamate receptor reflects the ability of nanosized compartments of the neuron to self-development (the transition to a new level of efficient synaptic transmission) using only their own resources. However, for maintaining the new state of spine resources from other compartments of the neuron are needed.

The synthesis of new proteins, occurring in the soma of the neuron, including AMPA receptors [23] is necessary to maintain long-term LTP. LTP in hippocampus is maintained for a long time, but not longer than 30–60 min if the synthesis of proteins is blockaded. To maintain the LTP in this initial period a post-translational modification of proteins and formation of protein-lipid vesicles in the vacuolar system [24] is required more than the protein synthesis. Maintaining the new level of transfer is accompanied by replacement of AMPA1/2 receptors on AMPA2/3 subtype [25]. The molecular network (PI2), and. interactive map is presented on the site (http://wwwmgs. bionet.nsc.ru/mgs/gnw/genenet/viewer/AMPA.html).

The basal stocks of AMPARs2/3 are in the vacuolar system, where they are included in the vesicles and delivered to the spines. The movement of proteins between compartments of vacuolar system (endoplasmic reticulum, Golgi apparatus, trans Golgi network, endosomes) and their delivery to the spine are mediated by small transport vesicles, branching off from a donor compartment and then fuse with an acceptor compartment [26]. The violation of vesicle assemblage induces a significant decline of the synaptic transmission efficiency in the first 20 min [27].

Functional and structural plasticity induced by external influences may lead to the synaptic activity shift to a non-physiological range. These mechanisms involved in learning and memory are balanced by another distinct form of neuronal modulation - homeostatic plasticity. One of the major cellular events underlying the expression of homeostatic regulation is the alteration of AMPARs accumulation and thus, synaptic strength [27]. These two forms of plasticity coexist to adapt to changing the external environment while maintaining the balance of neural activity within a physiological range. Neuronal networks use an array of homeostatic negative-feedback mechanisms that allow neurons to assess their activity and adjust accordingly, to restrain their activity within a physiological range [28, 29]. Different functional molecules and signaling cascades are involved in the expression of homeostatic up- or down regulation of synaptic activity and AMPAR expression.

4   Conclusion

Analysis of the cell as a functional system allows reconstructing the processes of the origination of integrative properties associated with the concepts of cognition, the functioning of neuron as the organism in the organism. PPI1, PPI2 net-works (fragment of cell maps) reflect the molecular systems functioning that required for neuron inclusion in the network of brain and maintenance of this communications with the cells of the network. Simulation of complex systems is necessary, because a multi-component molecular system of the nerve cell is perceived with difficulties without it. PPI1, PPI2 networks represent the basic set of the molecules and its functional inter-actions, which is required for translate the electrochemical signals into the following processes: inclusion of a neuron in the brain networks, recognition of synaptic activity patterns, converting the activation of glutamate receptors in changing the neurotrans-mission. The dendritic spines are dynamic structures. GeneNet computer system can reflect the organization of the protein-protein interaction networks in one of the possible states, because the temporal dynamics of the process can only be brought now in the form of a script. Perhaps PPI1, PPI2 networks can become the fragments of the complete electronic circuits of certain neuron type.

Acknowledgements. The work was supported by the basic project of fundamental research of RAS IV 35.1.5 and RFBR No. 15-29-04875-ofi_m and No. 17-04-01440-a.

References

1. Lin, L., Shen, S., Jiang, P., Sato, S., Davidson, B.L., Xing, Y.: Evolution of alternative splicing in primate brain transcriptomes. Hum. Mol. Genet. **19**(15), 2958–2973 (2010)
2. Penn, A.C., Balik, A., Wozny, C., Cais, O., Greger, I.H.: Activity-mediated AMPA receptor remodeling, driven by alternative splicing in the ligand-binding domain. Neuron **76**(3), 503–510 (2012)
3. Barbosa-Morais, N.L., Irimia, M., Pan, Q., et al.: The evolutionary landscape of alternative splicing in vertebrate species. Science **338**(6114), 1587–1593 (2012)
4. Sherrington, C.S.: Man on His Nature, p. 444. Cambridge University Press, Cambridge (1942)
5. Uexküll, J.V.: Theoretical Biology, p. 243. Harcourt, Brace and Co., New York (1926)
6. Anokhin, P.K.: Systems analysis of the integrative activity of the neuron. Pavlov. J. Biol. Sci. **19**(2), 43–101 (1974)
7. Fuster, J.M.: Cortex and memory: emergence of a new paradigm. J. Cogn. Neurosci. **21**(11), 2047–2072 (2009)
8. Pastalkova, E., Serrano, P., Pinkhasova, D., Wallace, E., Fenton, A.A., Sacktor, T.C.: Storage of spatial information by the maintenance mechanism of LTP. Science **313**(5790), 1141–1144 (2006)
9. Ratushnyak, A.S., Zapara, T.A.: Principles of cellular-molecular mechanisms underlying neuron functions. J. Integr. Neurosci. **8**(4), 453–469 (2009)
10. De Koninck, P., Schulman, H.: Sensitivity of CaM kinase II to the frequency of Ca^{2+} os-cillations. Science **279**(5348), 227–230 (1998)
11. Potter, W.B., O'Riordan, K.J., Barnett, D., Osting, S.M., Wagoner, M., Burger, C., Roopra, A.: Metabolic regulation of neuronal plasticity by the energy sensor AMPK. PLoS ONE **5**(2), e8996 (2010). doi:10.1371/journal.pone.0008996
12. Wang, G., Gilbert, J., Man, H.Y.: AMPA receptor trafficking in homeostatic synaptic plasticity: functional molecules and signaling cascades. Neural Plast. (2012). doi:10.1155/2012/825364
13. Aur, D., Mandar, J., Poznanski, R.R.: Computing by physical interaction in neurons. J. Integr. Neurosci. **10**(4), 413–422 (2011)
14. Cacha, L.A., Poznanski, R.R.: Associable representations as field of influence for dynamic cognitive processes. J. Integr. Neurosci. **10**(4), 423–437 (2011)
15. Red'ko, V.G., Mosalov, O.P., Prokhorov, D.V.: A model of evolution and learning. Neural Netw. **18**(5–6), 738–745 (2005)
16. Kolpakov, F.A., Ananko, E.A.: Interactive data input into the GeneNet database. Bioinformatics **15**(7–8), 713–714 (1999). (2005)
17. Craddock, T., Tuszynski, J.A., Hameroff, S.: Cytoskeletal signaling: is memory encoded in microtubule lattices by CaMKII phosphorylation? PLoS Comput. Biol. **8**(3), e1002421 (2012)
18. Valtschanoff, J.G., Weinberg, R.J.: Laminar organization of the NMDA receptor complex within the postsynaptic density. Neuroscience **21**(4), 1211–1217 (2001)
19. Tomita, S., Stein, V., Stocker, T.R., Nicoll, A., Bredt, D.S.: Bidirectional synaptic plasticity regulated by phosphorylation of stargazin-like TARPs. Neuron **45**(2), 269–277 (2005)
20. Choi, J., Ko, J., Park, E., Lee, J.R., Yoon, J., Lim, S., Kim, E.: Phosphorylation of stargazin by protein kinase A regulates its interaction with PSD-95. J. Biol. Chem. **277**(14), 2359–12363 (2002)
21. Carlisle, H.J., Manzerra, P., Marcora, E., Kennedy, M.B.: SynGAP regulates steady-state and activity-dependent phosphorylation of cofilin. J. Neurosci. **28**(50), 13673–13683 (2008)

22. Park, E., Na, M., Choi, J., Kim, S., Lee, J.R., Yoon, J., Park, D., Sheng, M., Kim, E.: The Shank family of postsynaptic density proteins interacts with and promotes synaptic accumulation of the beta PIX guanine nucleotide exchange factor for Rac1 and Cdc4. J. Biol. Chem. **278**(21), 19220–19229 (2003)
23. Lai, H.C., Jan, L.Y.: The distribution and targeting of neuronal voltage-gated ion channels. Nat. Rev. Neurosci. **7**(7), 548–562 (2006)
24. Vlachos, A., Maggio, N., Jedlicka, P.: Just in time for late-LTP: a mechanism for the role of pkmzeta in long-term memory. Commun. Integr. Biol. **1**(2), 190–191 (2008)
25. Newpher, T.M., Ehlers, M.D.: Glutamate receptor dynamics in dendritic microdomains. Neuron **58**(4), 472–497 (2008)
26. Hinners, I., Tooze, S.A.: Changing directions: clathrin-mediated transport between the golgi and endosomes. J. Cell Sci. **116**(Pt5), 763–771 (2003)
27. Malakchin, I.A., Proskura, A.L., Zapara, T.A., Ratushnyak, A.S.: Influence of transport vesicles assembly to preserve the effectiveness of the synaptic transmission Vestnik NGU (Russian), No. 101, pp. 14–20 (2012)
28. Davis, G.W.: Homeostatic control of neural activity: from phenomenology to molecular design. Ann. Rev. Neurosci. **29**, 307–323 (2006)
29. Hou, Q., Gilbert, J., Man, H.Y.: Homeostatic regulation of AMPA receptor trafficking and degradation by light-controlled single-synaptic activation. Neuron **72**(5), 806–818 (2011)

Effect of Persistent Sodium Current
on Neuronal Activity

E.Y. Smirnova[1,2(✉)], A.V. Zefirov[2,3], D.V. Amakhin[2],
and A.V. Chizhov[1,2]

[1] Ioffe Institute, Saint-Petersburg, Russia
elena.smirnova@mail.ioffe.ru
[2] Sechenov Institute of Evolutionary Physiology and Biochemistry
of the Russian Academy of Sciences, Saint-Petersburg, Russia
[3] Peter the Great St. Petersburg Polytechnic University, Saint-Petersburg, Russia

Abstract. In epilepsy, the number of persistent sodium (NaP) channels
increases. To study their effects on neuronal excitability we applied
dynamic-clamp (DC). We have revealed that NaP current decreases rheobase,
promotes depolarization block (DB) and changes membrane potential between
spikes. Bifurcation analysis of a Hodgkin-Huxley-like neuron reveals that NaP
current shifts saddle-node and Hopf bifurcations which correspond to the
rheobase and DB, in agreement with experiments. By shifting DB, NaP current
can make an antiepileptic effect via excitatory neurons.

1 Introduction

Experimental data demonstrate that the expression of such ionic channels as potassium
I_M [1], I_A [2], I_{BK} [3] and persistent sodium I_{NaP} [4–6] is changed in epilepsy. This
damage leads to development of chronic epilepsy. Agrawal et al. [5] have shown that in
pilocarpine model of epilepsy the expression of NaP channels increases. Here we test
whether this increase makes a compensatory effect or reverse. We suppose that such
effect might be provided by DB, the disruption of spike generation at too strong input
currents because of inactivation of sodium currents.

The main effects of NaP current have been revealed by Vervaeke et al. [7]. By
means of mathematical modeling and experiments in slices with DC, it has been found
that NaP channels decrease the gain of the rate-versus-current dependence, decrease the
rheobase, and improve regularity of spiking. The effect of NaP channels on DB has not
been considered, apparently due to technical limitations of the DC setup. According to
[8, 9], DB in pyramidal neurons could be one of the mechanisms of seizure cessation.
There are many factors affecting DB [9, 10]. Here we study the effects of NaP channels
on DB and other spiking characteristics by using DC and bifurcation analysis.

© Springer International Publishing AG 2018
B. Kryzhanovsky et al. (eds.), *Advances in Neural Computation, Machine Learning,
and Cognitive Research*, Studies in Computational Intelligence 736,
DOI 10.1007/978-3-319-66604-4_28

2 Methods

2.1 Experiment

Spike trains from pyramidal cells in the entorhinal cortex slices (deep layers) of the rat brain were recorded using the whole-cell patch-clamp technique in current-clamp and DC modes, see details in [10, 11] with conventional extracellular and potassium-gluconate-based pipette solutions. Hyperpolarizing and depolarizing current steps were injected for 1.5 s. In the DC mode, the NaP current was simulated according to the approximation from Vervaeke et al. [7]:

$$I_{NaP} = g_{NaP}\tilde{m}\left(V - V_{NaP}\right), \quad \frac{d\tilde{m}}{dt} = \frac{\tilde{m}_\infty - \tilde{m}}{\tau_{NaP}}, \quad \tilde{m}_\infty = \left(1 + \exp\left(\frac{V - V_{1/2}}{V_{slope}}\right)\right)^{-1}, \quad (1)$$

where $\tau_{NaP} = 1$ ms, $V_{NaP} = 30$ mV, $V_{1/2} = -51$ mV, $V_{slope} = -4.5$ mV fit to experimental data from [12]; g_{NaP} is variable maximal conductance.

2.2 Analysis of Experimental Data

Input resistance (R_{in}) was estimated as the gain of a linear approximation of a voltage-current relationship in the interval of injected currents from -60 to 0 pA. The membrane time constant (τ_m) was estimated from the response to -20 pA current step by an exponential fitting. The sag amplitude was measured from the response to -50 pA current step [13]. Other characteristics were: the spike threshold (Vth), defined as the membrane potential at the point with $dV/dt > 10$ mV/ms; the spike amplitude relative to the threshold level (RA); the spike half-width (HW) at the half-height; the after-spike hyperpolarization (AHP) after the third spike; its latency since spike t_{AHP}; the reverse interspike intervals IF_1 (between the first and second spikes), IF_3/IF_1 (relative third interval), and IF_n/IF_1 (relative last interval). Values of Vth, RA, HW were averaged across spikes. The frequency-current dependence was characterized by the gain (k), the rheobase, the current inducing maximal firing rate (I_{max}), and the limit current of DB. The rheobase was estimated with a step 10 pA, whereas DB with a step 50 pA. The firing rate (FR) was averaged during last 0.5 s of the current step in order to take into account only steady state regime. Comparison of the estimated parameters between cells was performed for the traces with half-maximal firing rates. Software programs were written in Wolfram Mathematica 10 and Delphi 7.

2.3 Hodgkin-Huxley-like Model of a Neuron

Bifurcation analysis was performed for a single compartment Hodgkin-Huxley-like neuron that includes the transient sodium current I_{Na}, the NaP current I_{NaP} (Eq. 1), the delayed rectifier potassium current I_K, and the leak currents of sodium, potassium and chloride ions. The model has been modified from [14] by assuming constant reversal potentials and adding the NaP current in the form of Eq. 1. The dynamics of the membrane potential, V, is described by the equations:

$$C\frac{dV}{dt} = I_{ext} - I_{Na} - I_K - I_L - I_{NaP},$$

$$I_{Na} = g_{Na}m^3h(V - V_{Na}), \qquad I_K = g_K n^4(V - V_K),$$

$$I_L = g_{Na_L}(V - V_{Na}) + g_{K_L}(V - V_K) + g_{Cl_L}(V - V_{Cl}),$$

$$\frac{dq}{dt} = \alpha_q(1 - q) - \beta_q q, \qquad q = m, h, n$$

(2)

Here I_{ext} is the external applied current; C is the membrane capacitance; V_j (j is either Na, K or Cl) are the reversal potentials calculated from the Nernst equation according to the composition of the perfusion and intracellular solutions used in our experiment (see Sect. 2.1); g_i are the maximal conductances, $i = Na$, K, Na_L, K_L, or Cl_L. The equations of opening and closing rate constants of the ion channel state transitions, taken from a pyramidal cell model [15]. The parameters were as follows: $C = 1$ μF/cm^2, $V_{Na} = 70$ mV, $V_K = -105$ mV, $V_{Cl} = -67$ mV, $g_{Na} = 30$ mS/cm^2, $g_K = 25$ mS/cm^2, $g_{Na_L} = 0.025$ mS/cm^2, $g_{K_L} = 0.05$ mS/cm^2, $g_{Cl_L} = 0.1$ mS/cm^2; g_{NaP} has been varied in numerical experiments. The bifurcation analysis has been performed in Matlab (using MatCont-toolbox).

3   Results

3.1   Dynamic-Clamp Study of the Influence of NaP Current

The NaP current was simulated in the DC mode with the approximation by Eq. 1. Figure 1 shows the effect of NaP current on the spike trains (A) and the frequency-current relationship (B) for a representative neuron. I_{NaP} decreases the rheobase (Rb) and the reversed first ISI, IF_1; promotes DB, and increases the spike frequency and the time parameter of AHP, t_{AHP}.

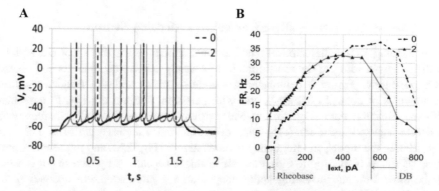

Fig. 1. Effect of NaP current simulated in DC on spiking of a representative neuron ($g_{NaP} = 2$ nS versus control, $g_{NaP} = 0$). **A.** Responses to the current step $I_{ext} = 40$ pA. **B.** Dependence of the firing rate (FR) on the external current I_{ext}.

Electrophysiological characteristics of neuronal spiking responses (Sect. 2.2) have been compared between groups of recordings in the control conditions ($g_{NaP} = 0$) and in addition of artificial NaP current ($g_{NaP} = 2$ nS), using paired t-test (n = 6, p < 0.05). The statistically significant effects are given in Table 1. Other parameters that were not affected significantly are (mean ± SEM): $R_{in} = 172 \pm 39$ MOm; $\tau_m = 20 \pm 2$ ms; $I_{max} = 550 \pm 80$ pA; $k = 0.11 \pm 0.02$ Hz; $TFS = 264 \pm 50$ ms; $IF_n/IF_1 = 0.60 \pm 0.11$; $RA_m = 68 \pm 2$ mV; $Vth_m = -41 \pm 1$ mV; $HW_m = 1.1 \pm 0.3$ ms; $AHP = 9.5 \pm 1.2$ mV; $Sag = 2.4 \pm 0.5\%$; $V_{rest} = -66 \pm 2$ mV. These results are consistent with modeling and the experiments with DC from [7]. The decrease of the DB limit due to NaP current is a novel observation. As noted in Sect. 1, DB may play a crucial role in the cessation of seizure, thus the reported increase of NaP channel density [5] may be a compensatory reaction of a neuron through promotion of DB.

Table 1. Effects of NaP current on neuronal firing characteristics

Parameter	$g_{NaP} = 0$	$g_{NaP} = 2$ nS	p-value
IF_1, Hz	27 ± 3	21 ± 2	0.0186
Rb, pA	62 ± 21	23 ± 14	0.0282
DB, pA	650 ± 93	560 ± 110	0.0313
IF_3/IF_1	0.73 ± 0.09	0.95 ± 0.15	0.0499
t_{AHP}, ms	12 ± 1	19 ± 2	0.0116

Values are means ± SEM. p-value characterizes the paired t-test for n = 6, p < 0.05. $g_{NaP} = 2$ nS.

The NaP current increases t_{AHP}. This effect explains the discrepancy of the majority of mathematical models of a single neuron with experiments. Typical experimental spike trains show t_{AHP} as large as 10÷30 ms, whereas in models it is commonly less than 5 ms. Because AHP is associated with the action of calcium-dependent potassium currents, the discrepancy might be wrongly explained by inacurate approximation of potassium channels or electrotonic effects. Alternatively, it may be caused by not accounting of NaP currents.

3.2   Effect of Persistent-Sodium Current in a Modeled Neuron

Understanding of mechanisms of NaP channel influence on the rheobase and DB implies understanding of the transitions from a steady state to oscillations and reverse, correspondingly. Figure 2 demonstrates the spike trains and the frequency-current relationships in control conditions and with persistent sodium current. In agreement with experimental data (Fig. 1), the NaP current increases the spike frequency for moderate injected currents (Fig. 2A), decreases the rheobase and the DB limit (Fig. 2B). In the model, the frequency is much higher (Fig. 2A), and the repolarization phase of spikes is also quite different, which might be caused by two major reasons: (i) the slow currents providing spike adaptation are not taken into account in the considered model taken from [14]; (ii) contribution of fast potassium channels into the repolarization phase is overestimated, because of inaccuracy of Hodgkin-Huxley

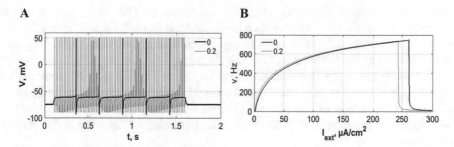

Fig. 2. Effect of NaP current in the model. **A.** Responses to the current step I_{ext} = 1.86 μA/cm^2 in control conditions and with additional I_{NaP} (g_{NaP} = 0.2 mS/cm^2). **B.** Dependence of spike frequency (FR) on the external current I_{ext}.

formalism applied for sodium channels (known as "window current" problem [16] that exposes as a lack of pure sodium spikes in Hodgkin-Huxley model). Nevertheless, the mentioned shortcomings of the model are result in quantitative but not qualitative inconsistency with our experiments.

Bifurcation analysis performed for variable external current I_{ext} (Fig. 3) has revealed two bifurcations known to occur in such a model: the supercritical Andronov-Hopf (H) and saddle-node on invariant circle (SN) bifurcations. In control conditions with g_{NaP} = 0 (Fig. 3A), the SN-bifurcation corresponds to the transition of the system from silent to spiking state. The bifurcation occurs around I_{ext} = 1.86 A/cm^2. Between SN- and H- bifurcations, the spike amplitude decreases with I_{ext}. The

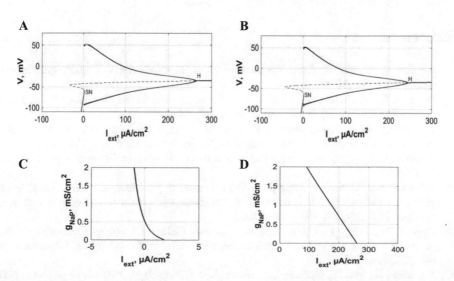

Fig. 3. Bifurcation analysis. **A.** Diagram for V as function of I_{ext} in control and with additional NaP current, g_{NaP} = 0.2 mS/cm^2. **B.** Solid line between SN and H is the limit cycle. **C, D.** Two-parameter bifurcation diagrams (g_{NaP}; I_{ext}) at SN- (C) and H- (D) bifurcations. The domain of spiking is at right in C and left in D.

H-bifurcation around $I_{ext} = 262$ μA/cm^2 corresponds to DB. With I_{NaP} (Fig. 2B, $g_{NaP} = 0.2$ mS/cm^2), the SN-bifurcation occurs around $I_{ext} = 0.72$ μA/cm^2 and H does around $I_{ext} = 244.72$ μA/cm^2. Again, in agreement with experiments (Sect. 3.1), NaP current decreases the rheobase and the DB limit.

The two-parameter bifurcation diagrams shown in Fig. 3C reveals saturation for large g_{NaP} in the case of SN-bifurcation. Figure 3D shows that the DB limit decreases with g_{NaP}. Comparison of Fig. 3C with D reveals that the NaP current stronger affects DB. Hence, the main effect of NaP curret is the narrowing of the range of the external current that evokes firing.

4   Conclusion

In experiments with DC we have revealed some effects of NaP current on neuronal excitability: decrease of the rheobase, promotion of DB, increase of spike frequency and changes of the voltage profile between spikes. Mathematical modeling with a Hodgkin-Huxley-like neuron has approved the effects qualitatively and revealed quantitative discrepancies in the input-output functions. Bifurcation analysis has shown that the range of the external current able to evoke spikes is limited by the saddle-node and Hopf bifurcations. The NaP current shifts the bifurcations and narrows the range of excitation. Related to excitatory neurons, the effects produce an antiepileptic effect. Opposite, NaP current in inhibitory interneurons might produce a proepileptic effect.

Acknowledgments. This work was supported by the Russian Science Foundation (Project 16-15-10201).

References

1. Kohling, R., Wolfart, J.: Potassium channels in epilepsy. Cold Spring Harb Perspect Med. **6** (a022871), 1–24 (2016)
2. Bernard, C., Anderson, A., Poolos, N., Becker, A., Beck, H., Johnson, D.: Acquired dendritic channelopathy in epilepsy. Science **305**, 532–535 (2004)
3. Pacheco Otalora, L.F., Hernandez, E.F., et al.: Down-regulation of BK channel expression in the pilocarpine model of temporal lobe epilepsy. Brain Res. **1200**, 116–131 (2008)
4. Stafstrom, C.E.: Persistent sodium current and its role in epilepsy. Epilepsy Curr. **7**(1), 15–22 (2007)
5. Agrawal, N., Alonso, A., Ragsdale, D.S.: Increased persistent sodium currents in rat entorhinal cortex layer V neurons in a post-status epilepticus model of temporal lobe epilepsy. Epilepsia. **44**, 1601–1604 (2003)
6. Royeck, M., Kelly, T., Opitz, T., et al.: Downregulation of spermine augments dendritic persistent sodium currents and synaptic integration after status epilepticus. J. Neurosci. **35** (46), 15240–15253 (2015)
7. Vervaeke, K., Hu, H., Graham, L.J., Storm, J.F.: Contrasting effects of the persistent Na+ current on neuronal excitability and spike timing. Neuron **49**, 257–270 (2005)
8. Jirsa, V.K., Stacey, W.C., Quilichini, P.P., Ivanov, A.I., Bernard, C.: On the nature of seizure dynamics. Brain. **137**, 2210–2230 (2014)

9. Yekhlef, L., Breschi, G.L., Taverna, S.: Optogenetic activation of VGLUT2-expressing excitatory neurons blocks epileptic seizure-like activity in the mouse entorhinal cortex. Sci. Rep. **7** (2017). Article 43230

10. Smirnova, E.Y., Zaitsev, A.V., Kim, K.K., Chizhov, A.V.: The domain of neuronal firing on a plane of input current and conductance. J. Comput. Neurosci. **39**(2), 217–233 (2015)

11. Amakhin, D.V., Ergina, J.L., Chizhov, A.V., Zaitsev, A.V.: Synaptic conductances during interictal discharges in pyramidal neurons of rat entorhinal cortex. Front. Cell. Neurosci. **10**, 1–15 (2016). Article 233

12. Hu, H., Vervaeke, K., Storm, J.F.: Two forms of electrical resonance at theta frequencies, generated by M-current, h-current and persistent Na+ current in rat hippocampal pyramidal cells. J. Physiol. **545**, 783–805 (2002)

13. Zaitsev, A.V., Povysheva, N.V., Gonzalez-Burgos, G., Lewis, D.A.: Electrophysiological classes of layer 2/3 pyramidal cells in monkey prefrontal cortex. J. Neurophysiol. **108**, 595–609 (2012)

14. Wei, Y., Ullah, G., Ingram, J., Schiff, S.J.: Oxygen and seizure dynamics: II computational modeling. J Neurophysiol. **112**, 213–223 (2014)

15. Gloveli, T., Dugladze, T., Schmitz, D., Heinemann, U.: Properties of entorhinal deep layer neurons projecting to the rat dentate gyrus. Eur. J. Neurosci. **13**, 413–420 (2001)

16. Borg-Graham L.J.: Interpretations of data and mechanisms for hippocampal pyramidal cell models. In: Cerebral Cortex, pp 19–138. Springer (1999)

Printed in the United States
By Bookmasters